OXFORD CHEMISTRY F

Physical Chemistry Editor
RICHARD G. COMPTON
University of Oxford

Founding/Organic Editor
STEPHEN G. DAVIES
University of Oxford

Inorganic Chemistry Edi
JOHN EVANS
University of Southampto

CW00547807

Organonitrogen Chemistry

Patrick D. Bailey

Professor of Organic Chemistry at Heriot-Watt University, Edinburgh

Keith M. Morgan

Research Associate in Organic Chemistry at Heriot-Watt University, Edinburgh

OXFORD

UNIVERSITY PRESS

*This book has been printed digitally and produced in a standard specification
in order to ensure its continuing availability*

OXFORD
UNIVERSITY PRESS

Great Clarendon Street, Oxford OX2 6DP

Oxford University Press is a department of the University of Oxford.
It furthers the University's objective of excellence in research, scholarship,
and education by publishing worldwide in

Oxford New York

Auckland   Cape Town Dar es Salaam   Hong Kong   Karachi
Kuala Lumpur Madrid Melbourne Mexico City   Nairobi
New Delhi Shanghai Taipei Toronto
With offices in
Argentina Austria Brazil Chile Czech Republic France Greece
Guatemala Hungary Italy Japan South Korea Poland Portugal
Singapore Switzerland Thailand Turkey Ukraine Vietnam

Oxford is a registered trade mark of Oxford University Press
in the UK and in certain other countries

Published in the United States
by Oxford University Press Inc., New York

ISBN 0-19-855775-2

Printed and bound by CPI Antony Rowe, Eastbourne

Contents

Series Editor's Foreword

Organonitrogen compounds are amongst the most ubiquitous in chemistry and nature. The chemical properties of organonitrogen functional groups span the whole gamut of reactivity and mechanism and hence the topic provides an ideal focus for teaching organic chemistry: surprisingly this opportunity is rarely if ever exploited fully in standard texts.

Oxford Chemistry Primers have been designed to provide concise introductions relevant to all students of chemistry, and only contain the essential material that would usually be covered in an 8–10 lecture course. This Organic Chemistry Primer provides an excellent and easy to read account of the universal topic of organonitrogen chemistry. It provides the basis for both specific courses and for the supplementary study of most aspects of organic chemistry conventionally dealt with in complementary ways throughout all chemistry courses.

This primer will be of interest to apprentice and master chemist alike.

Stephen G. Davies
The Dyson Perrins Laboratory, University of Oxford

Preface

Nitrogen plays a pivotal role in organic chemistry, and there are a huge number of organonitrogen functional groups in natural products, man-made drugs, and new materials. Organonitrogen chemistry is so diverse that coverage in most organic textbooks is very fragmented. However, in this OUP primer we have brought together the features and reactions of organonitrogen functional groups, and the key principles which underly virtually all of the chemistry have been emphasized in short introductions to each of the three Parts of the book. We hope this will simplify matters for the newcomer at early undergraduate level, and may provide a clearer understanding (or useful revision) for those who are more experienced.

Organonitrogen chemistry is still delivering a wealth of fascinating research results, and the authors are currently studying aspects of asymmetric synthesis, heterocyclic synthesis, and the chemistry of unusual peptides. It is our continued fascination with nitrogen chemistry, and support from friends and family, that has spurred us on to complete this book. In particular, P. D. B. would like to thank his family for their support (Judy, Hannah, Thomas, and 'Junior'), and his research group for their inspiration and enthusiasm, and K. M. M. would like to dedicate this book to his grandmother and grandfather.

Edinburgh   P. D. B.
October 1995   K. M. M.

Introduction

Nitrogen constitutes 80 per cent of the air we breathe, and is the most abundant gas on earth. Despite the lack of reactivity of the element itself, nature has found several ways of incorporating nitrogen into the molecules of life. This is fortunate indeed, for it is impossible to imagine the evolution of life in which organonitrogen chemistry does not play a crucial role.

But what is so special about the presence of nitrogen in organic compounds? The answer lies in a combination of factors.

	C	N	O
Number of bonds (when neutral)	4	3	2
Number of lone pairs	0	1	2

Firstly, neutral nitrogen is trivalent, and this opens up a whole range of functional groups and structural variants that are inaccessible to its divalent neighbour oxygen. Secondly, trivalent nitrogen possesses a lone pair of electrons; this not only influences its chemistry, but also allows nitrogen to utilize these unpaired electrons in additional functional groups (e.g. ammonium ions and nitro groups).

So let's have a quick look at some common organonitrogen functional groups.

Saturated	Unsaturated	N–N and N–O bonded
CH$_3$CH$_2$ — NH$_2$ (amine)	CH$_3$ – CH= NMe (imine)	CH$_3$CH$_2$ – NH- NH$_2$ (hydrazine)
CH$_3$CH$_2$ — $\bar{\text{N}}$H (metal amide)	CH$_2$ = CH- NMe$_2$ (enamine)	CH$_3$CH$_2$ —N=$\overset{+}{\text{N}}$=$\bar{\text{N}}$ (azide)
CH$_3$CH$_2$ — $\overset{+}{\text{N}}$H$_3$ (alkylammonium cation)	CH$_3$ NH$_2$ (amide)	CH$_3$CH$_2$ — $\overset{+}{\text{N}}$Me$_2$ (N-oxide)
CH$_3$CH$_2$ — $\overset{+}{\text{N}}$Me$_3$ (quaternary ammonium ion)	MeC≡N (nitrile)	CH$_3$CH$_2$ —N (nitro)

The examples listed above have been divided into three groups, whose chemistry will be covered in more detail in Parts A, B, and C of this book.

In Part A, we will look at simple saturated amines. These can be protonated or alkylated to give ammonium cations, or deprotonated to generate anions. Although this chemistry is relatively straightforward, the

Trivalent nitrogen functional groups are neutral on nitrogen.

Tetravalent nitrogen must share its lone pair of electrons, and thus carries a positive charge. e.g.

R— NMe$_2$ →(MeI)→ R— $\overset{+}{\text{N}}$Me$_3$

R—N=O →([O])→ R—$\overset{+}{\text{N}}$ (with O and O$^-$)

Amides contain the functional group:

'**Metal amides**', on the other hand, are the salts of amines: e.g. sodamide (NaNH$_2$), lithium diisopropylamide (LiNPr$^i_2$).

There are three acceptable representations of the **nitro** group:

— $\overset{+}{\text{N}}$ (O, O$^-$)   — N (O, O)   — $\overset{+}{\text{N}}$ (O, O)

We will use the first valence bond representation for this and related functional groups, as this simplifies mechanistic arrow pushing.

properties and synthesis of simple amines cover important aspects of organonitrogen chemistry that relate to many other functional groups.

Much more varied chemistry becomes available when unsaturation is present, and these compounds are covered in Part B. Imines, enamines, amides, and nitriles are representative of the wide range of structural units in which the nitrogen forms part of a π-conjugated system.

A further range of functional groups is accessible when N–N and N–O bonding occurs. These groups can be either saturated (e.g. *N*-oxides) or unsaturated (e.g. nitro compounds), and a bewildering array of combinations is known. A selection of these functional groups is discussed in Part C.

As if that were not enough, there is a range of heterocyclic nitrogen compounds that adds further interest. Those that are aromatic (e.g. pyridine, imidazole) are covered in another OUP primer (*Aromatic Heterocyclic Chemistry*, by D. T. Davies) but non-aromatic ones like morpholine and pyrrolidine behave pretty well like their acyclic counterparts, and such cyclic compounds will be discussed in this book.

But before we start to look at the chemistry of organonitrogen compounds in earnest, it is perhaps worth reminding ourselves of the importance of these compounds both in nature and in the industrialized world.

At the core of all life, controlling the very form and function of all organisms, is the genetic code. This is composed of a (nitrogen-free) polymer backbone of ribose sugars linked by phosphate groups. The code itself is contained within the sequence of nitrogen-rich bases (B^n); their hydrogen-bonding potential holds the DNA double helix together, and is also utilized when the code (in triplets of bases) is transformed into a chain of amino acids.

pyridine   imidazole

morpholine   pyrrolidine

base joined to backbone here

NH$_2$

Adenine (A)

Thymine (T)

A and T are paired by H-bonding

NH$_2$

Cytosine (C)

Guanine (G)

C and G are paired by H-bonding

DNA

If B^1=A, B^2=T, and B^3=C, the triplet of bases encodes alanine (Ala)

Me

= Ala

The DNA-controlled condensation of amino acids generates peptides and proteins that carry out most of the functions of life. Peptides (< 50 amino acids) are primarily chemical messengers, whilst proteins (> 50 amino acids) provide much of the structure of skin and bone, help in the functioning of cells, and act as receptors, enzymes, and transporters.

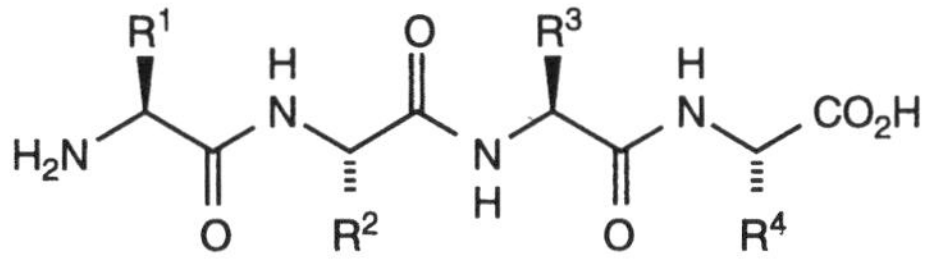

a peptide 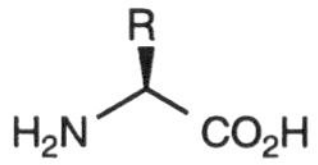   an amino acid

Many plants contain alkaloids which are simply nitrogen-containing natural products whose toxicity was harnessed in classical herbal medicines, and whose biological properties have prompted their use as 'lead structures' in modern drug design.

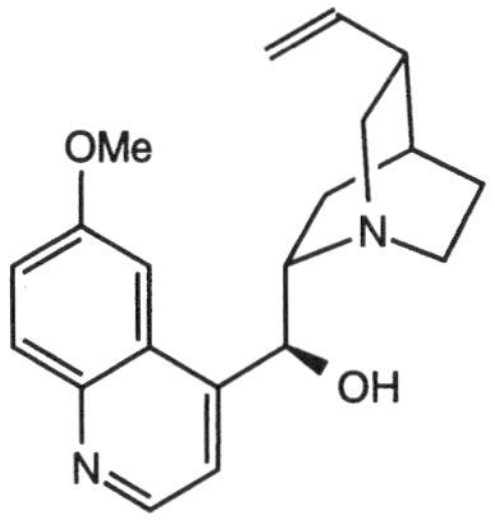

quinine
(natural anti-malarial)

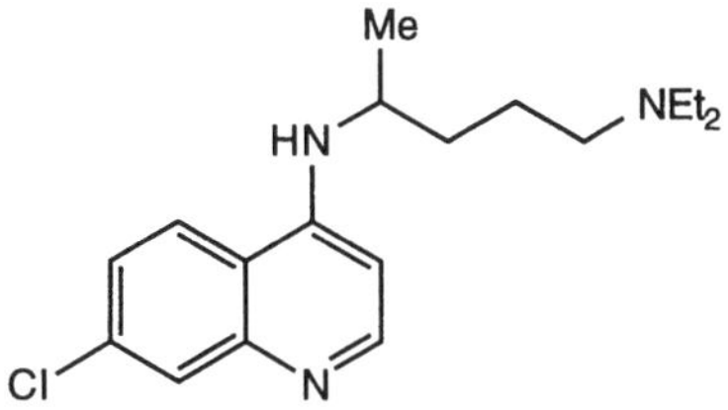

chloroquine
(synthetic anti-malarial)

Organonitrogen compounds also feature strongly in man-made materials, such as the synthetic fibres and dyes that led to the development of the modern chemical industry.

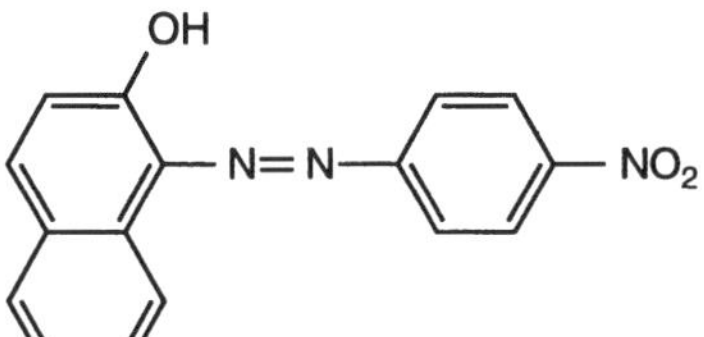

para red (a dye)

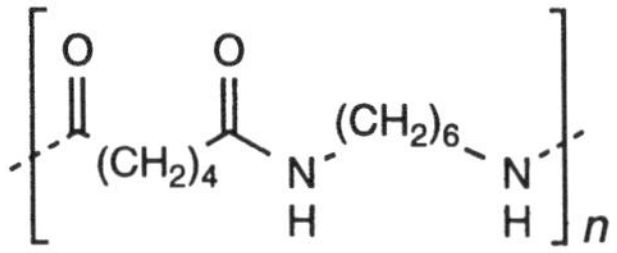

nylon-6,6 (a synthetic fibre)

So organonitrogen chemistry really is central to modern life, relating to natural products like DNA, peptides, proteins, alkaloids...and to man-made pharmaceuticals, fibres, and dyes. Nitrogen is probably the most important element in organic chemistry—with the possible exception of carbon!

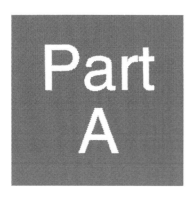

Saturated nitrogen compounds

There are only a few (closely related) nitrogen functional groups that are fully saturated:

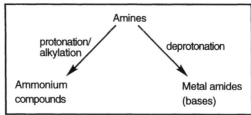

We will include in this section unsaturated derivatives of the above functional groups if they essentially leave the chemistry of the nitrogen unaltered (e.g. $PhCH_2NH_2$). Here are the factors that control the chemistry and synthesis of these compounds:

- strong C–N bond
- slightly polarized C–N bond
- lone pair on neutral nitrogen
- nitrogen will reluctantly accept a negative charge
- nitrogen is sp^3-hybridized

Typical bond strengths are:

C–C	345–355 kJ mol^{-1}
C–N	290–315 kJ mol^{-1}
C–O	355–380 kJ mol^{-1}

Taking these points in turn:

(a) An average C–N σ-bond is almost as strong as a typical C–C single bond. Therefore, a C–N single bond is very hard to break—usually the nitrogen needs to be charged.

(b) The small difference in electronegativity of carbon and nitrogen leads to slight polarization of the C–N bond.

(c) Critically, neutral nitrogen has a lone pair of electrons. This has two important consequences. Firstly, it opens up mechanistic routes that are unavailable to carbon.

e.g. $CH_3\overset{..}{N}H_2$ + $CH_3–I$ ⟶ $CH_3\overset{+}{N}H_2–CH_3$ $\xrightarrow{-H^+}$ $(CH_3)_2NH$

cf. CH_3CH_3 + $CH_3–I$ ⟶̸

Secondly, mechanisms involving tetravalent N^+ are prevalent; nitrogen can hold a positive charge and still have eight outer electrons. All carbon cations (carbenium ions) have only six electrons, and are consequently less stable than their nitrogen counterparts. Oxygen cations are also less stable than their nitrogen equivalents even though both can possess a full octet of electrons. This is because oxygen is more electronegative than nitrogen (e.g. $MeNH_2$ is essentially fully protonated to $MeNH_3^+$ at pH 7 whereas MeOH requires pH $\ll$ 1 to be protonated as $MeOH_2^+$).

(d)   The electronegativity argument works the other way where anions are concerned. Thus, nitrogen anions are very powerful bases; for example, Et_2NH is only deprotonated at pH > 23! In contrast, EtOH is deprotonated at pH 16.

(e)   Finally, just a reminder that most saturated organonitrogen compounds possess tetrahedral sp^3-hybridized nitrogen.

Amines		
Primary	1°	R^1NH_2
Secondary	2°	R^1R^2NH
Tertiary	3°	$R^1R^2R^3N$
Quaternary	4°	$R^1R^2R^3R^4N^+$

Amine natural products include:

HO, HO — dopamine (with NH₂)

dopamine

HO, HO — adrenaline (with OH and N-Me)

adrenaline

indolizidine

lycopodine

histrionicotoxin

1 Amines

Saturated amines are the simplest organonitrogen compounds, but they are important because they occur widely in nature, are often used as building blocks for more complex compounds, and are used as co-reagents in many organic reactions. Their reactivity is dependent on the nitrogen lone pair, which facilitates reactions in which the nitrogen acts either as a **nucleophile** or as a **base**.

A **nucleophile** is something that attacks electron-deficient carbon (an electrophile)

A **base** is something that attacks electron-deficient hydrogen (i.e. picks up a proton)

Throughout this book, we will use:
Nu⁻ for nucleophile (– meaning lone pair of electrons)
B⁻ for base (– meaning lone pair of electrons)
E⁺ for electrophile (+ meaning can accept pair of electrons)

The interplay between nucleophilicity and basicity is a subtle one that we will return to shortly. But we need to begin by looking at the very simplest of reactions of amines—when they act as bases. We will start by using ethylamine, $EtNH_2$, as a specific example.

1.1 Basicity

This is simply a measure of where the following equilibrium lies:

$$B^- + H_2O \rightleftharpoons BH + OH^-$$

For a typical amine like ethylamine, this equilibrium is:

$$EtNH_2 + H_2O \rightleftharpoons Et\overset{+}{N}H_3 + OH^-$$

The fact that this equilibrium lies over to the right tells us that aqueous solutions of amines (like aqueous ammonia itself) are alkaline.

For this equilibrium, the equilibrium constant can be expressed as:

$$K = \frac{[\overset{+}{Et\,N\,H_3}][OH^-]}{[EtNH_2][H_2O]}$$

Of course the concentration of water is always (virtually) the same, so this is simply incorporated into the constant K. This new constant is defined as the basicity constant, K_b. So:

$$K_b = K[H_2O] = \frac{[\overset{+}{Et\,N\,H_3}][OH^-]}{[EtNH_2]}$$

What is most useful is to have a feel for when the free amine and its protonated ammonium ion have equal concentrations. It turns out that K_b is about 10^{-5}, so when $[\overset{+}{Et\,N\,H_3}] = [EtNH_2]$:

$$K_b = 10^{-5} = [OH^-]$$

This still isn't a very useful relationship. But pH is a much more usable idea, and $[OH^-]$ can be converted into pH because:

$$[H^+][OH^-] = 10^{-14}$$
$$\therefore \quad \log_{10}\{[H^+][OH^-]\} = \log_{10} 10^{-14}$$
$$\therefore \quad \log_{10}[H^+] + \log_{10}[OH^-] = -14$$

and $\quad pH = -\log_{10}[H^+] \quad$ and $\quad pOH = -\log_{10}[OH^-]$

$$\therefore \quad pH + pOH = +14$$
$$\therefore \quad pH = 14 - pOH$$

so when $[OH^-] = 10^{-5}$

$$pOH = -\log_{10} 10^{-5} = 5$$
$$\therefore \quad pH = 14 - pOH = 14 - 5 = 9$$

What this tells us is that typical amines are half-ionized at about pH 9. At neutral pH 7, when the H^+ concentration is 100 times higher (as pH is a $-\log_{10}$ scale), 100 times as much amine will be protonated as unprotonated.

In practice, working with basicity constants is a real nuisance; it's easier to work with the equilibrium in reverse. So:

$$Et\,\overset{+}{N}\,H_3 + H_2O \rightleftharpoons EtNH_2 + H_3O^+$$

This is the much more familiar equilibrium for acidity, for which:

$$K = \frac{[EtNH_2][H_3O^+]}{[Et\,\overset{+}{N}\,H_3][H_2O]}$$

and $\qquad K_a = K[H_2O] = \dfrac{[EtNH_2][H_3O^+]}{[Et\overset{+}{N}H_3]}$

K_a for typical amines is about 10^{-9}, so when $[EtNH_2] = [Et\overset{+}{N}H_3]$:

$$[H_3O^+] = 10^{-9}$$

$$\therefore \quad pH = -\log_{10}[H_3O^+] = -\log_{10} 10^{-9} = 9$$

Of course, this is the same answer that we got before. But it is a much quicker way to see that the amine is half-ionized at pH 9. The proportion of protonated amine will go up by a factor of 10 for every unit reduction in pH; conversely, we need to have the pH > 11 before > 99% of an aqueous solution of an amine exists as free RNH_2 rather than as the protonated form.

Basicity of amines
- Most amines give **alkaline** aqueous solutions
- *Basicity* is usually measured using the **acidity** constant
  $pK_a = -\log_{10}[K_a]$
- Typical aliphatic amines have a $pK_a \approx 9$
- When the pH < 9, most of the amine is protonated
- For **more basic** amines, the pK_a is **higher**

All simple amines have pK_a values of about 9, but the exact value is modified by the nature of the substituents. Here are a few pK_a values to give you a feel for this:

Amine	Structure	pK_a
Ammonia	NH_3	9.26
Ethylamine	$EtNH_2$	10.75
Diethylamine	Et_2NH	10.98
Triethylamine	Et_3N	10.64
Piperidine	NH	11.20
Aniline	$PhNH_2$	4.58
Pyridine	N	5.23

Electron-donating groups clearly stabilize the positive charge on the nitrogen of protonated amines, reducing K_a and therefore raising the pK_a. The cyclic amines piperidine and pyrrolidine are perfectly normal, but pyridine and pyrrole are affected by aromaticity, and aniline by the conjugated benzene ring:

$$R_3\ddot{N} + \overset{+}{H} \rightleftharpoons \overset{+}{R_3NH}$$

If R is an electron-donating group then this equilibrium lies over to the right. The equilibrium lies over to the left if R can conjugate with the nitrogen lone pair.

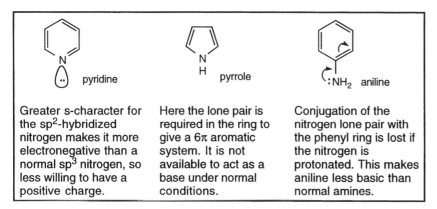

pyridine	pyrrole	:NH₂ aniline
Greater s-character for the sp²-hybridized nitrogen makes it more electronegative than a normal sp³ nitrogen, so less willing to have a positive charge.	Here the lone pair is required in the ring to give a 6π aromatic system. It is not available to act as a base under normal conditions.	Conjugation of the nitrogen lone pair with the phenyl ring is lost if the nitrogen is protonated. This makes aniline less basic than normal amines.

Hydrogen bonding to water stabilizes aqueous ammonium ions, and facilitates hydrogen exchange in amines.

It is important to remember that gaining and losing protons from an amine is kinetically a very easy process. The exchange of protons is very rapid (i.e. thousands of times a second); the pK_a tells us the **average** ionization that we can expect for an amine at a given pH.

$$R-\overset{H}{\underset{H}{\overset{+}{N}}}-H\cdots\cdots:OH_2$$

Our ability to control the protonation of amines using pH can be neatly exploited for their purification. At pH < 7, amines are usually protonated and since water is highly polar, the protonated amines are generally water soluble. Non-ionized organic impurities can now be washed out using an organic solvent. Finally, the amino component can be recovered by raising the pH (to > 11), and extracting the free amine into an organic phase.
E.g. mixture of PhCO₂H, PhCO₂Et, and PhCH₂NH₂:

Q. How might you separate the following using acid and/or base extractions? (Hint! pH control.)

PhNH₂, EtNH₂, HO₂CCH₂NH₂

1.2 Nucleophilicity

Nitrogen follows carbon in the periodic table, and it forms strong C–N bonds. In general, primary and secondary amines react readily with electron-deficient (i.e. electrophilic) carbon.

Typical bond strengths (kJ mol⁻¹):

C–N 305	C–C 347
C=N 615	C=C 611
C≡N 891	C≡C 837

For example:

$$Et\overset{..}{N}H_2 \quad \curvearrowright CH_3 \overset{\frown}{-}I \longrightarrow Et-\underset{H_2}{\overset{+}{N}}-Me \ + \ I^- \xrightarrow{-H^+} Et-\underset{H}{\overset{+}{N}}-Me$$

$$\underset{EtNH_2}{\overset{H_3C}{\diagup}}\overset{H_3C}{\diagdown}C=O \longrightarrow Et-\underset{H_2}{\overset{+}{\underset{|}{N}}}-\underset{\underset{CH_3}{|}}{\overset{\overset{CH_3}{|}}{C}}-O^- \xrightarrow{-H_2O} \underset{CH_3}{\overset{Et}{\diagdown}}N=\underset{\diagdown CH_3}{\overset{\diagup CH_3}{C}}$$

However, with tertiary amines, the extra bulk and the inability to generate a neutral product (no protons to lose) reduce the nucleophilicity enormously. So tertiary amines (e.g. Et_3N) usually act as bases rather than nucleophiles. It is perhaps useful to see where amines come in the scale of nucleophilicity, basicity, and leaving group ability, when compared with other species:

Nucleophilicity:

$I^- > Et_2NH > CN^- > MeO^- > Br^- > NH_3 > $ pyridine $ > Cl^- > MeCO_2^- > F^- > MeOH$

Leaving group ability:

$I^- \approx Br^- \approx H_2O > Cl^- > MeCO_2^- > CN^- > NH_3 > Et_2NH > OH^- > MeO^- > NH_2^-$

Basicity (and approximate pK_a):

$NH_2^- > OH^- > MeO^- > Et_2NH > NH_3 > CN^- > $ pyridine $ > MeCO_2^- > F^- > H_2O > MeOH > Cl^- > Br^- > I^-$

36	16	15	10	9	9	5	5	3	−2	−2	−7	−9	−10

A further subtlety is the idea of 'hard' and 'soft' nucleophiles and bases (see introduction to Part B).

From this interplay of nucleophilicity and basicity, it is clear that amines are unusually versatile. In general, the nucleophilic nature of nitrogen can be totally quenched by the addition of acid; protonation of the lone pair of electrons completely removes its ability to attack electrophilic carbon.

1.3 Reactions of amines

With alkyl halides. The standard S_N2 reaction takes place readily:

$$Et\overset{..}{N}H_2 \quad \diagdown Me\overset{\frown}{-}I \longrightarrow Et-\underset{H_2}{\overset{+}{N}}-Me \ + \ I^- \xrightarrow{-HI} Et-\underset{H}{\overset{+}{N}}-Me$$

This reaction does not usually stop here, however, as the product is generally a more powerful nucleophile than the starting material:

$$Et-\underset{\underset{Me}{|}}{\overset{..}{N}}H \quad \diagdown Me\overset{\frown}{-}I \longrightarrow Et-\underset{\underset{Me}{|}}{\overset{+}{N}}H-Me \ + \ I^- \xrightarrow{-HI} EtNMe_2$$

And a third reaction can also occur, to form the ammonium salt:

$$Et\overset{..}{N}Me_2 \quad \diagdown Me\overset{\frown}{-}I \longrightarrow \overset{+}{Et}NMe_3 \ I^-$$

Nucleophilicity of different amines

Reactions are usually fastest for:

$$2° \quad > 1° \quad > 3°$$
$$R^1R^2NH > R^1NH_2 > R^1R^2R^3N$$

This is because nitrogen is **more** nucleophilic as it becomes **more** electron rich (+I effect of alkyl groups), but **less** nucleophilic as its environment becomes more crowded.

The synthesis of amines is discussed in Chapter 4.

Alkylation of amines with methyl halides is actually rather hard to control (unless you want the quaternary ammonium salt) because the methyl group is so small. With larger alkylating groups, it is often quite easy to stop the alkylation at the tertiary amine.

The reactivity of amines is also strongly influenced by **solvent effects**—changing from a polar protic solvent (e.g. MeOH) to an aprotic non-polar solvent (e.g. PhH) can modify the reactivity of amines enormously.

Alkylations need a base

At each stage of the reaction of ethylamine with iodomethane, an equivalent of HI is generated. This is a strong acid, that would protonate the amino reactant...and tie up its lone pair. Such reactions grind to a halt before completion unless a base is added to neutralize the acid or pick up the proton. Typical bases are:

$$NaHCO_3 \qquad K_2CO_3 \qquad NEt_3 \qquad pyridine$$
$$\text{(neutralize acid)} \qquad\qquad \text{(pick up proton)}$$

Inorganic bases are conveniently filtered off after the reaction. Organic bases are usually tertiary amines, emphasizing that amines have dual functionality as nucleophiles (substrate) and bases (co-reagent). But remember, in the case of iodomethane as the carbon electrophile, tertiary amines can themselves be alkylated. The use of nitrogen bases is discussed further in Chapter 3.

With carboxylic acid derivatives. Amines can again act as nucleophiles, forming amides:

e.g.

The amides produced are themselves **non-basic** and **poor** nucleophiles because the nitrogen lone pair is tied up in conjugation with the carbonyl group (see Chapter 7).

e.g.

The two-step reaction mechanism is often abbreviated as shown, with the double-headed arrow to oxygen indicating the formation and then collapse of the tetrahedral intermediate.

Note that HCl is formed when an amine reacts with an acid chloride. The reaction will not reach completion unless a base is also present to stop protonation of the starting amine...a second equivalent of the amine can be used as the base.

In general, amines react with most carboxylic acid derivatives to form amides, with the speed of the reaction being controlled by the reactivity of the carbonyl compound:

X = Cl	Acid chloride	Very fast
X = (anhydride)	Anhydride	Fast (very convenient)
X = OEt	Ester	Very slow

With free carboxylic acids, no reaction occurs; the acid simply protonates the amine:

$$EtNH_2 + CH_3CO_2H \rightleftharpoons Et\overset{+}{N}H_3 + CH_3CO_2^-$$

With aldehydes and ketones. As with carboxylic acid derivatives, primary and secondary amines can attack the electron-deficient carbonyl carbon. The products are imines or enamines which result from overall loss of water (see Chapters 5 and 6).

Q. Write a mechanism for these reactions, which are acid catalysed.

Cleaving C–N bonds. With C and N being adjacent to each other in the periodic table, the C–N single bond is very strong. Nitrogen is a poor leaving group, and the small difference in electronegativity makes the carbon only slightly electron deficient. Saturated amines are therefore pretty stable compounds. Only *N*-benzyl amines can be cleaved under relatively mild conditions—hydrogenolysis over a palladium catalyst usually does it. If this fails, adding an acid catalyst, or using the more active palladium hydroxide [Pd(OH)$_2$–C; Pearlman's catalyst] usually does the trick. Sodium in liquid ammonia also removes benzyl.

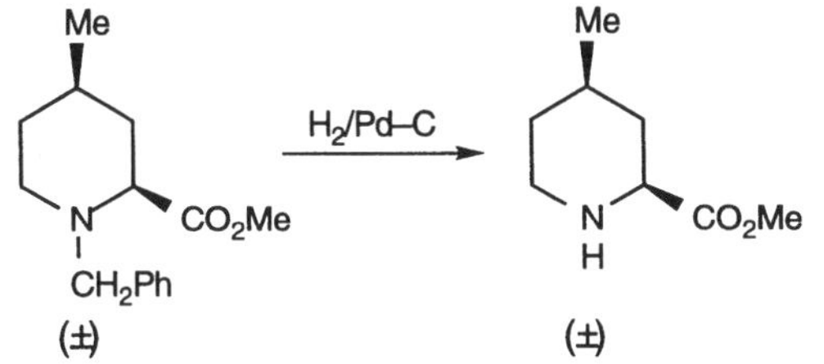

In the example shown above, the benzyl group was actually used as a protecting group, so that the nitrogen would not act as a nucleophile during steps involved in its synthesis—remember that 3° amines usually only act as **bases**, whilst 2° amines are good **nucleophiles**.

> **α-Methylbenzylamine**
>
> When α-methylbenzyl is used in place of benzyl, there is a stereogenic centre present which often can be used to control the absolute stereochemistry of reactions involving the nitrogen. The starting material for such syntheses is usually α-methylbenzylamine, which is cheap, and available as the (*R*)- or (*S*)-isomer. Once the α-methylbenzyl **auxiliary** has directed the **stereochemistry** and is no longer required as a **protecting group**, it can be removed just like the normal benzyl group.
>
> e.g.
>
> 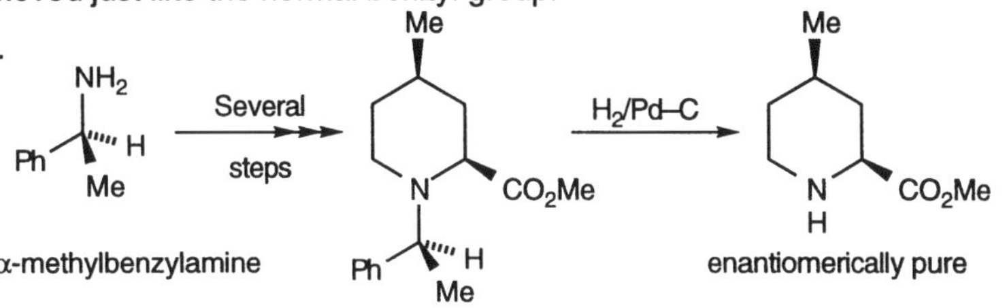
>

Reactions with nitrous acid. These will be discussed further in Chapter 13, but the key transformations are:

1° Amine: $RNH_2 \xrightarrow{HNO_2(aq)} [R-\overset{+}{N}\equiv N]$ (unstable) $\xrightarrow{H_2O} ROH$

2° Amine: $R_2NH \xrightarrow{HNO_2(aq)} R_2N\text{–}N\text{=}O$ (yellow)

3° Amine: $R_3N \xrightarrow{HNO_2(aq)}$ No reaction

Aromatic Amine: $ArNH_2 \xrightarrow{HNO_2(aq)} [Ar-\overset{+}{N}\equiv N]$ (stable) $\xrightarrow{Cu^{I}X} ArX$

These reactions are the classical tests for 1°, 2°, and 3° amines.

This is the Sandmeyer reaction.

1.4 Summary

- Amines are bases, which are readily protonated (typical pK_a around 9).
- Amines are powerful nucleophiles, and react with:
  alkyl halides;
  carboxylic acid derivatives;
  aldehydes and ketones;
  nitrous acid.

2 Ammonium compounds

Ammonium compounds have the general formula $R^1R^2R^3R^4N^+$, where R^n can be alkyl, aryl, or hydrogen. But they fall into two distinct classes. Firstly, those in which one or more of the R groups is hydrogen; these ammonium cations are formed by simple protonation of an amine, and their formation is readily reversed.

$$R^1R^2R^3N + H_3O^+ \rightleftharpoons R^1R^2R^3NH^+ + H_2O$$

This is actually just another way of writing the K_b equilibrium discussed in Chapter 1 (just try adding OH^- to both sides). Importantly, when base is added to protonated amines, the ammonium ion just loses the proton to regenerate the free amine, and no other reactions occur.

The second class of ammonium ions are called quaternary ammonium ions. For these cations, all four R^n groups in $R^1R^2R^3R^4N^+$ are alkyl or aryl. Crucially, these cations are not in equilibrium with a neutral amine, and we will look at their properties, how they are made, and their key reactions.

2.1 Properties of quaternary ammonium ions

Quaternary ammonium compounds have a number of important properties, of which we will look at three:

(i)   chirality;
(ii) solubility in solvents;
(iii) phase transfer catalysis.

Chirality

Amines possessing three different groups attached to nitrogen are potentially chiral. In practice, such compounds invert their stereochemistry thousands of times per second (the so-called 'umbrella' effect):

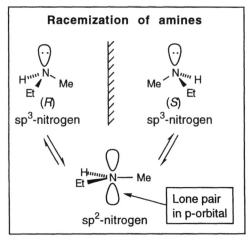

One example of a naturally occurring ammonium compound is **acetylcholine**, an essential neurotransmitter.

It is released when a nerve ending receives an impulse. By altering the potential across the synapse, acetylcholine allows transmission of the impulse to the next nerve cell. Once the impulse has been transmitted, the acetylcholine is removed by enzymatic hydrolysis of the ester group. Organophosphate 'nerve gases' inhibit this enzymatic process leading to a series of continuous nerve impulses, paralysis, and death.

Some amines **do** have a stereogenic nitrogen that does not invert, but only when rings prevent inversion. e.g.

(A)   (B)

(C)

Trögers base (C) was the first chiral amine to be resolved that did **not** possess any chiral carbon atoms.

Protonated amines are also potentially chiral, but this time the enantiomers cannot be resolved because proton exchange always occurs during attempted resolution—and the tertiary amine can then racemize.

For quaternary ammonium ions, racemization cannot readily occur via either mechanism, so these compounds exist as separable enantiomers, provided that all four R groups are different.

If the racemic ammonium cation is allowed to form a salt with an enantiomerically pure carboxylic acid (e.g. (*R*)-2-phenylpropanoic acid, (*R*)-Ph–CHMe–CO$_2$H), then the two ion pairs ((*R*)-ammonium ion/(*R*)-acid and (*S*)-ammonium ion/(*R*)-acid) are diastereoisomers, and should have slightly different properties. With a bit of luck, one ion pair might be persuaded to precipitate from a solvent in which the other pair is soluble. Chiral ammonium compounds are primarily of academic interest, but chiral phase transfer catalysts (see below) can be used to generate optically active products in some reactions.

This idea is the basis of **resolution** whereby mixtures are separated at the end of a racemic synthesis. Prior to the development of modern asymmetric methods, this was the most commonly employed route to single enantiomers of chiral synthetic targets.

Solubility in solvents

Quaternary ammonium ions are schizophrenic! The charged ammonium part prefers polar solvents like water (it is hydrophilic or water-loving); the alkyl/aryl R groups are hydrocarbons that prefer non-polar organic solvents like benzene, ethoxyethane, dichloromethane (which are hydrophobic or water-hating). The result is pretty obvious—in a two-phase water/organic solvent system, quaternary ammonium compounds tend to partition appreciably into **both** phases. If the R groups are generally small, or have some hydrophilic functional groups, then the cation is mainly present in the aqueous phase. Conversely, if the R groups are large and hydrophobic, then the cation favours the organic phase.

Solubility of quaternary ammonium ions

Me—N⁺(Me)(Me)(Me) I⁻

Me(CH₂)₃—N⁺((CH₂)₃Me)((CH₂)₃Me)(Me(CH₂)₃) I⁻

Tetramethylammonium iodide
Positive nitrogen dominates
Water soluble (slightly soluble in Et₂O)

Tetrabutylammonium iodide
Alkyl groups dominate
Et₂O soluble (slightly soluble in water)

Phase transfer catalysis

So far, we have pretended that the quaternary ammonium species can exist on its own, and we have tried to look at its properties in isolation from any counter-ion. But whether as a solid or in solution, ammonium ions must be associated with an equivalent number of counter-ions. The **solubility** properties of the ammonium part can be used to control the solubility of the counter-ion.

This trick is most widely used when inorganic nucleophiles are being used with organic substrates in an organic solvent. If a quaternary ammonium compound is added, some of the cation will enter the organic phase (see previous section). The cation has to carry a counter-ion with it in order to maintain electrical neutrality. The positive and negative ions needn't travel around as tightly bound salt pairs. In fact, bulky groups on the ammonium ion can shield it from any associated anion and make the anion more nucleophilic. For example, the rate of displacement of a halogen (X) with CN^- can be greatly enhanced by the addition of a small amount of phase transfer catalyst.

Some (increasingly bulky) examples of commercially available **phase transfer catalysts**:

$Et_4\overset{+}{N}\,Br^-$, $Bu_4\overset{+}{N}\,Br^-$, $Bn\overset{+}{N}\,Me_3\,Br^-$

$[CH_3(CH_2)_{11}]_4\overset{+}{N}\,Br^-$

In the example below NaCN is the source of CN^-, but concentration in the organic phase is low. The quaternary ammonium salt:

$Bn\overset{+}{N}\,Me_3\,Br^-$

is the phase transfer catalyst, increasing the concentration of CN^- in the organic phase, and thus increasing the rate of the reaction.

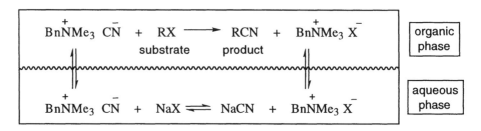

2.2 Synthesis of quaternary ammonium ions

Starting from a tertiary amine the key step is usually an S_N2 reaction.

$$\text{e.g. } Me_3N\colon + PhCH_2Br \longrightarrow Me_3\overset{+}{N}CH_2Ph + Br^-$$

If all four alkyl groups on the ammonium ion are the same, then the synthesis is particularly simple—just add excess alkylating agent to ammonia (and a base like $NaHCO_3$ to mop up the protons); successive S_N2 reactions should ultimately yield the desired product.

$$\text{e.g. } NH_3 + EtBr \xrightarrow{\;NaHCO_3\;} \overset{+}{N}Et_4 + Br^-$$
$$\text{(excess)}$$

On the other hand, when there is a mixture of R groups, the synthesis of a suitable tertiary amine ($R^1R^2R^3N$) for the final alkylation can be a major problem—methods for the synthesis of amines are discussed in Chapter 4. If one or more of the R groups are aryl, they cannot usually be introduced at the final step (only strongly electron-deficient aryl halides are susceptible to nucleophile attack)—so aryl groups would normally be introduced earlier in the synthesis.

2.3 Reactions

Considering that they are charged, quaternary ammonium ions are perhaps surprisingly stable; but remember that the C–N bond is strong, and there is no nitrogen lone pair available at which reactions can easily occur. In fact, the only two important reactions are elimination and reduction.

Elimination

Quaternary ammonium compounds will usually undergo elimination reactions when treated with strong base.

$$\underset{\substack{\beta \quad\;\; \alpha}}{\overset{H}{\diagdown}\;CH_2-CH_2} \xrightarrow{\;Me\overset{-}{O}Na^+\;} MeOH + CH_2\!=\!CH_2 + NMe_3$$

Ammonium ions lacking a β-hydrogen atom (e.g. $PhCH_2\overset{+}{N}Me_3$) cannot undergo this elimination, and are generally stable to base—they are useful phase transfer catalysts if harsh reaction conditions are required.

Because the C–N bond is strong, elimination reactions of quaternary ammonium ions virtually never follow the E1 pathway (i.e. unimolecular loss of a leaving group, in this case a tertiary amine, as the first step of the reaction). In almost all cases, E2 kinetics are followed, in which the base and substrate are both involved in the rate-determining step.

Elimination reactions—a reminder

$$\underset{X}{\overset{H}{\underset{\diagdown}{\text{CR}_2 - \text{CR}_2}}} \quad \xrightarrow{\text{Base}} \quad H^+ \; + \; \text{CR}_2 = \text{CR}_2 \; + \; X^-$$

E 1
- X^- drops off as the first, rate-determining step (r.d.s.)
- Rate α [substrate]

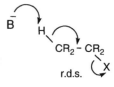

E1$_{cb}$
- H^+ drops off as the first step; this is **always** reversible, so the subsequent loss of X^- is the r.d.s.
- Reaction is first-order in [$\bar{\text{C}}\text{R}_2$–CR_2–X], the conjugate base (cb) of the substrate
- These reactions follow second-order kinetics
- Rate α [substrate] x [base]

E 2
- H^+ and X^- leave in a single base-induced step
- The energy paid for breaking **two** bonds is offset by the simultaneous formation of the new π-bond
- This requires both the H–C and C–X bonds to be co-planar; *anti*-periplanar is usually the lowest energy conformation (i.e. not *syn*):

H–C–C–X *anti*-periplanar   H–C–C–X *syn*-periplanar

steric repulsion

Quite strong bases are usually needed to get these elimination reactions to work. An alkoxide (e.g. sodium methoxide in methanol) or strong heating in aqueous sodium hydroxide is usually effective. Heating with moist silver oxide also works well, as will be discussed below.

The most important feature of these elimination reactions is the direction of elimination when there is a choice.

Elimination from quaternary ammonium compounds

Given a choice of two or more alkenes, the least substituted one predominates—this is Hoffmann's rule

e.g.

$CH_2 = CH_2$

$Me_2C = CMe_2$

$\gg$

Hoffmann's observations are in stark contrast to most other E2 eliminations (e.g. from alkyl halides, perhaps the commonest examples). Because the double bond forms **during** an E2 reaction, we might expect the more stable alkene to be formed preferentially—and this is the case for alkyl halides...hence Saytzeff's rule that eliminations generate the most substituted alkene. But two factors drive the Hoffmann elimination of quaternary ammonium ions towards less substituted alkenes:

- The reaction has 'a hint of' El_{cb} about it. As the C–N bond is strong, and the nitrogen has a powerful –I effect, the C–H bond **starts** to break first. The most acidic proton has the **least** +I alkyl groups attached, and loss of this proton therefore generates a less substituted alkene.

- The transition state is very crowded—but less so if the base attacks the proton attached to the least substituted carbon.

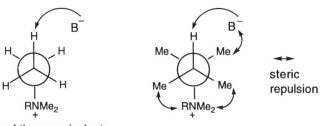

One of three equivalent low energy conformations

High energy conformation

steric repulsion

Even with a small base, the bulky ammonium ion must be in a high energy conformation if elimination is to take place (*anti*-periplanar). Even the smallest quaternary ammonium group (i.e. $-NMe_3^+$) is roughly the size of a tertiary butyl group (i.e. $-CMe_3$).

In the following example the importance of *anti*-periplanar groups in an E2 elimination is obvious!

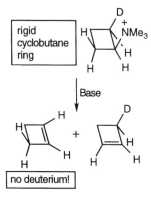

However, some compounds that undergo Hoffmann elimination cannot attain an *anti* conformation. Obviously something interesting is happening! A neat experiment to investigate this is:

Base

no deuterium!

So, E2 elimination can also take place if the angle between the leaving group and the proton is 0°. This (rare) elimination is therefore *syn*-periplanar.

E2 Elimination with large base

In general, really large bases cannot approach the more hindered proton that would lead to the more substituted alkene—Hoffmann elimination then occurs (less substituted alkene formed).

	MeONa/MeOH			Bu^t–COK/Bu^t–OH		
(Bu, Cl)	10	:	1	1	:	10
(Bu, $\overset{+}{N}Me_3$)	1	:	10	1	:	10

Hoffmann was a brilliant chemist, who utilized his elimination reaction to great effect for the structural analysis of naturally occurring alkaloids, long before spectroscopic methods were to dominate such work. He would fully methylate any nitrogens to the quaternary ammonium ions with methyl iodide (Hoffmann exhaustive methylation), then trigger elimination by heating with moist silver oxide (Ag$_2$O). The silver reacted with the iodide counter-ion, generating the hydroxide salt (with the OH$^-$ acting as the base) which, on heating, would undergo elimination. This elimination sequence would be repeated until no nitrogen was left, and the resulting (poly)alkene could then be analysed further.

e.g.

Reduction

Reduction of quaternary ammonium compounds can be achieved with powerful reducing agents such as LiAlH$_4$ or Na/NH$_3$(liq). However, yields are only modest and mixed products are formed unless all four substituents on nitrogen are the same.

More selective is the Emde reaction, using sodium amalgam (Na–Hg) in water. Only unsaturated groups are cleaved, and this can allow selective reduction of aryl–N bonds as demonstrated in the following example.

e.g.

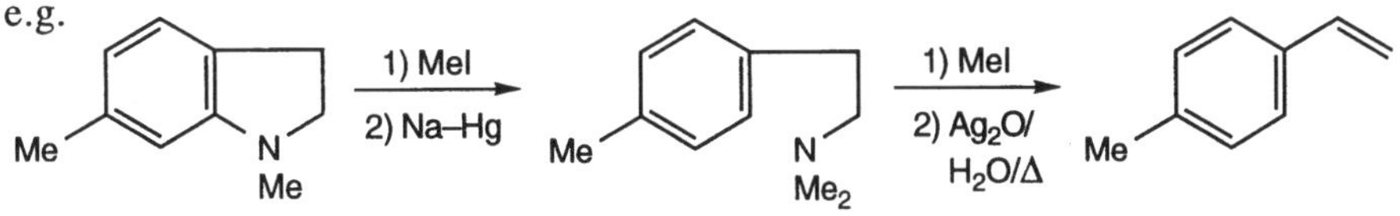

We can see from the above scheme that the reduction and elimination of quaternary ammonium salts can yield valuable structural information, and this has been used in the classical structure determinations of alkaloids.

2.4 Summary

- Quaternary ammonium ions can be used as phase transfer catalysts to aid transport of counter-anions into an organic phase.
- They are usually prepared by fully alkylating an amine with excess alkyl halide.
- Reaction with base leads to E2 elimination, giving the least substituted alkene (Hoffmann elimination).
- Powerful reducing agents can reduce quaternary ammonium compounds to tertiary amines.

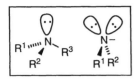

3   Nitrogen bases

Nitrogen compounds are widely employed as bases in organic synthesis. They can play the role of a **passive** base by mopping up excess protons to ensure a reaction doesn't become too acidic, or they can act as **active** bases by actually removing a proton from an organic substrate. The base we choose really depends on the type of reaction for which it is needed.

3.1 Types of nitrogen base

In general, primary and secondary amines aren't used, because they can act as nucleophiles as well as bases. The two commonest types of nitrogen base are tertiary amines ($R^1R^2R^3N$), and metal amides ($M^+NR_2^-$). We can grade the following commonly used examples by their basicity, and you may find the table below very useful. Remember, the pK_a for basicity is the pH at which exactly half of the base is protonated.

e.g. $Et_3\overset{+}{N}H \underset{}{\overset{K_a}{\rightleftharpoons}} Et_3N + H^+$ (Basicity)

e.g. $Et_2NH \underset{}{\overset{K_a}{\rightleftharpoons}} Et_2\overset{-}{N} + H^+$

Base	Abbreviation	Basicity (pK_a)
Pyridine	py	5.2
Ammonia	NH_3	9.2
Trimethylamine	TMA	9.8
Triethylamine	TEA	11.0
Hydroxide	OH^-	15.7
Methoxide	MeO^-	16.0
t-Butoxide	Bu^tO^-	18.0
Ethylamine 'amide'	$EtNH^-$	≈ 35

3.2 Reactions employing nitrogen bases

We can identify three types of reaction that utilize a base:

(a) *Reactions which would fail to reach completion if a base was absent.*
   e.g.

$$CH_3CH_2NH_2 \;+\; \underset{O}{\overset{CH_3}{\underset{\|}{\bigvee}}} Cl \;\longrightarrow\; \underset{O}{\overset{H}{CH_3CH_2N}}\underset{\|}{\bigvee}CH_3 \;+\; HCl$$

The HCl generated in this reaction would protonate the amine starting material, and the alkylation would soon grind to a halt at half completion. A base is simply needed to mop up the HCl. Bases like pyridine, NEt_3, or the more hindered Hünig's base are widely used.

(b) *Reactions involving concerted removal of a proton in the rate-determining step.*
Good examples are many E2 eliminations.

e.g.

Ph—CH(CH₂Br)CH₂ $\longrightarrow$ Ph—CH=CH₂ + BH + Br⁻

In this case, the better the base, the faster the reaction. But the proton doesn't always have to be stripped off the carbon prior to the elimination ($E1_{cb}$), so a moderately powerful base is often adequate. Common examples would be DBN, DBU, or DABCO. Pyridine is also commonly employed; although only a modest base it can be used as a solvent, heated conveniently to reflux if necessary, and readily removed later on a rotary evaporator.

(c) *When a proton needs removing to give an anionic intermediate.*
The most powerful nitrogen bases are needed.

e.g.

$CH_3CO_2Bu^t$ $\xrightarrow{\text{Base}}$ $\bar{C}H_2$–C(=O)OBu^t $\longleftrightarrow$ CH_2=C(O⁻)OBu^t $\xrightarrow{PhCH_2Br}$ CO_2Bu^t–CH₂–PhCH₂

The most widely used base for this type of deprotonation is lithium diisopropylamide (LDA), generated by treating diisopropylamine with butyllithium.

$(Me_2CH)_2NH + BuLi \longrightarrow (Me_2CH)_2\bar{N} Li^+ + BuH$

In order to avoid side-reactions occurring whilst the anion is formed, deprotonations are usually conducted at low temperatures (typically −78°C). Quenching of the anion (with benzyl bromide in the above example) is usually also carried out at low temperature, and then the mixture is allowed to warm up to ensure complete reaction. The advantage of LDA over even stronger bases like butyllithium is that LDA is usually too hindered to act as a nucleophile:

Hünig's base or diisopropylethylamine (DIPEA):

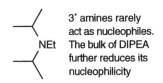

3° amines rarely act as nucleophiles. The bulk of DIPEA further reduces its nucleophilicity

1,5-diazabicyclo[4.3.0]non-5-ene (DBN):

1,8-diazabicyclo[5.4.0]undec-7-ene (DBU):

1,4-diazabicyclo[2.2.2]octane (DABCO):

Q. Why are DBN and DBU stronger bases than simple 3° amines ?

LDA removes the **kinetic** (least hindered) proton. For example:

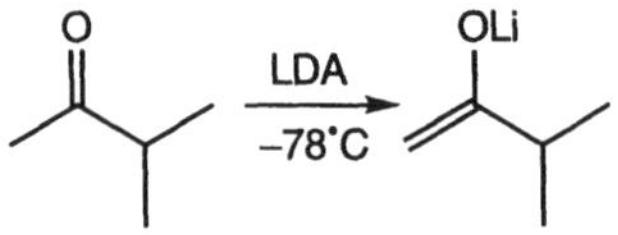

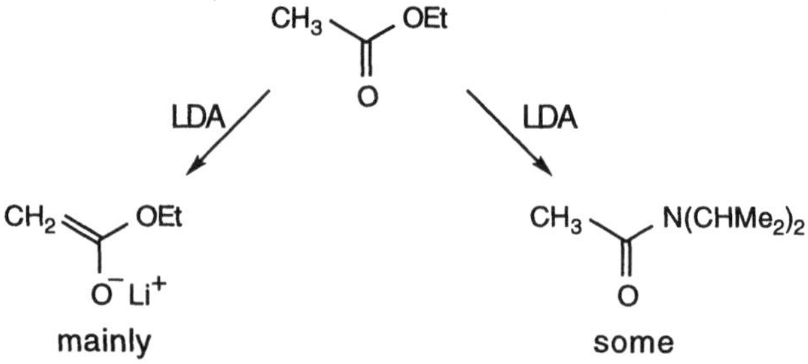

If LDA does lead to unwanted nucleophilic attack, the even more hindered lithium hexamethyldisilazide (LHMDS, $Li^+ \bar{N}(SiMe_3)_2$) can be employed.

Basicity of nitrogen versus oxygen

Nitrogen is less electronegative than oxygen, and this explains its use as a base. In contrast, ROH is essentially non-basic:

i.e. $ROH + H^+ \rightleftharpoons R\overset{+}{O}H_2$

but $R_3N + H^+ \longrightarrow R_3\overset{+}{N}H$

Conversely, it is much harder to deprotonate an amine than an alcohol (again due to electronegativity effects); but once an amine **is** deprotonated, the anion is much more reactive than its alkoxy counterpart:

i.e. $RO^- + H^+ \rightleftharpoons ROH$

but $R_2N^- + H^+ \longrightarrow R_2NH$

3.3 Summary

- Good bases for mopping up protons are NEt_3, DIPEA, or pyridine.
- To help a proton leave as part of a concerted reaction, try one of the diazabicyclo amines (DBN, DBU, DABCO).
- To strip off a proton, use LDA at $-78°C$.

4 Synthesis of amines

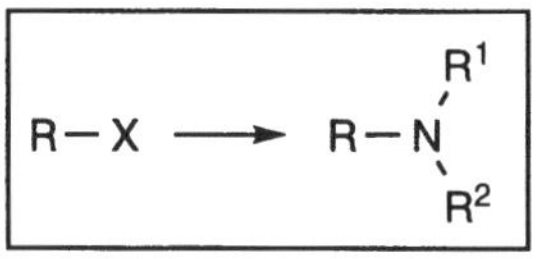

$$R-X \longrightarrow R-N\begin{smallmatrix} R^1 \\ \\ R^2 \end{smallmatrix}$$

Amines are accessible from three main sources. Some (e.g. ethylamine, pyridine) are directly available from natural sources or as major industrial products—these are the basic amine building blocks. Others can be synthesized by the introduction of the amino group to a suitable precursor. Finally, simple amines can be alkylated (directly or indirectly) to yield more complex amines.

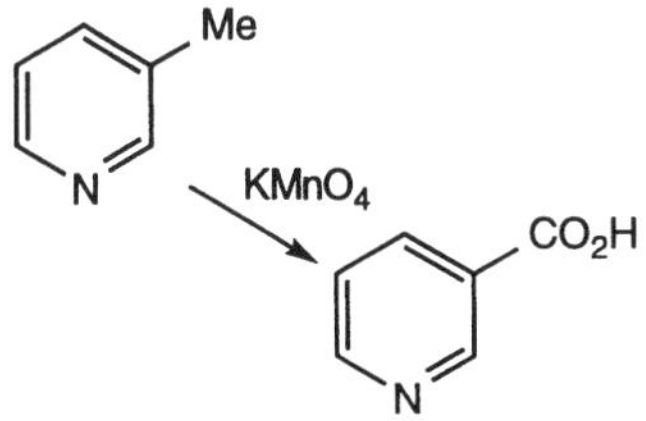

There are two types of reaction which are widely used for preparing or interconverting amines—substitution and reduction.

4.1 Synthesis of amines via substitution reactions using amines or ammonia

The simplest route to amines is displacement of a leaving group by ammonia. However, this is rarely of practical use, because further substitution reactions almost invariably occur:

e.g.

$$MeI \xrightarrow{NH_3} MeNH_2 \xrightarrow{MeI} Me_2NH \longrightarrow Me_3N \longrightarrow Me_4N^+ I^-$$
$$+ HI \qquad\qquad + HI \qquad\qquad + HI$$

So, if methylamine ($MeNH_2$) or dimethylamine (Me_2NH) was required, the other amino by-products would have to be removed. The order of reactivity of amines is controlled by both inductive and steric effects.

i.e. Reactivity of $Me_2NH > MeNH_2 > NH_3 > Me_3N$

There are two ways of getting the above reaction (i.e. methylation of ammonia) to yield largely a single product:

Use a large excess of NH₃. In the above example, $MeNH_2$ would then be the main product, because the MeI will only rarely encounter any nucleophile other than ammonia. Of course, once all the MeI has been used up, no further reaction is possible. These reactions use the excess ammonia as a base (see below). This tactic is rarely practicable in organic synthesis, so special reagents that act like a mononucleophilic ammonia have been devised (see the next section).

Use a large excess of MeI. This will generate the ammonium salt, $Me_4\overset{+}{N}\overset{-}{I}$, and this reaction forms the basis of the Hoffmann exhaustive methylation/elimination procedure for degrading amines. In general, any amine can be fully methylated using an excess of MeI, although a base must be added to make sure that the liberated HI doesn't simply protonate the amine, and slow down any further alkylation.

Pyridine is found in coal tar along with various methyl-pyridines. Oxidation of 3-methylpyridine gives the vitamin niacin:

e.g.

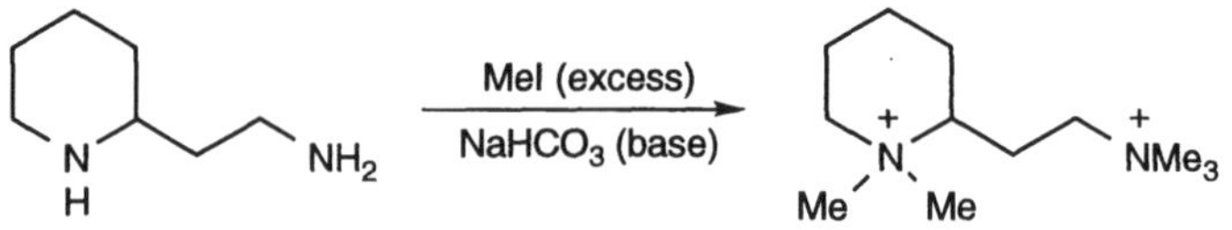

When amines are alkylated with groups larger than methyl, steric crowding makes the quaternization step very slow, and it is then often quite easy to stop the reaction at the tertiary amine stage.

e.g.

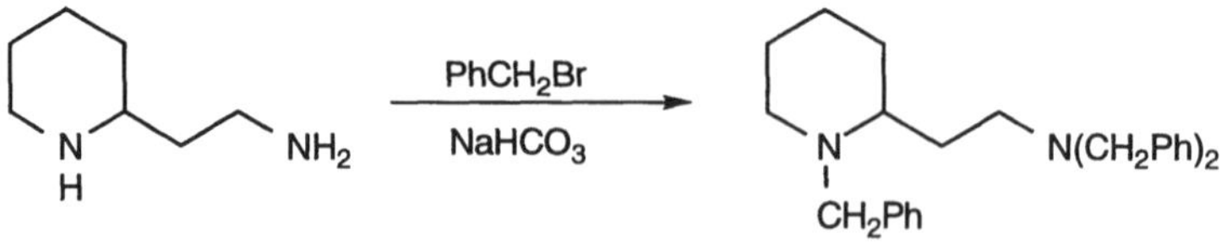

4.2 Using mononucleophilic 'ammonia'

There are several of these reagents now available, and they all require a second chemical step in order to liberate the free amino group:

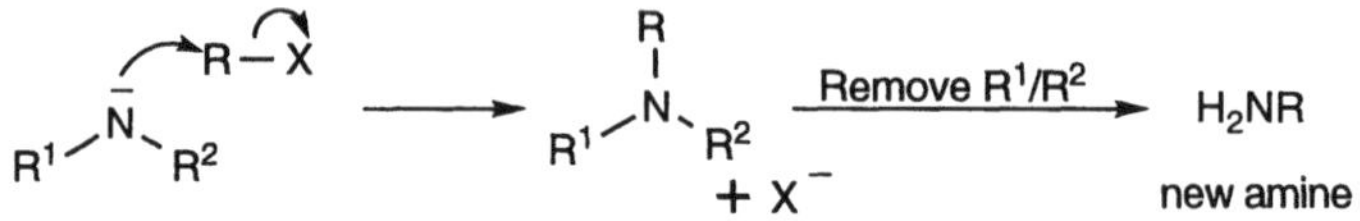

Here are five widely used methods:

(a) The classical Gabriel method

React the alkyl halide with potassium phthalimide, then reflux with OH⁻ (aq) or with H₂N–NH₂. The Gabriel reagent (potassium phthalimide) can be made from readily available phthalic acid or phthalic anhydride, but it is usually bought ready for use.

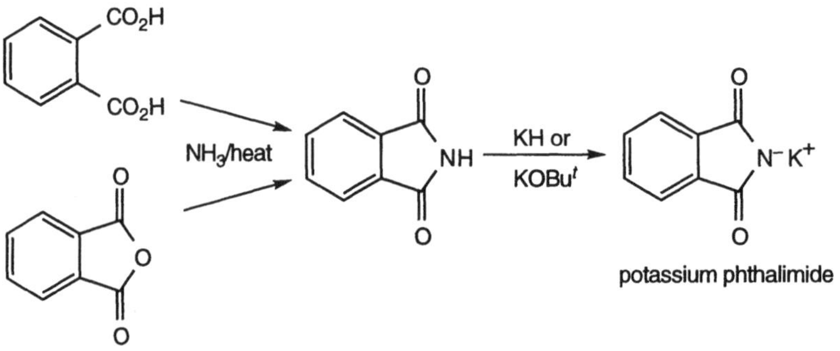

potassium phthalimide

Because the negative charge is delocalized on to the adjacent carbonyl oxygens, the Gabriel reagent acts as a nucleophile rather than a base, and readily displaces leaving groups from organic molecules. Once the nitrogen has acted as a nucleophile, it is essentially inert since its lone pair is

The **Gabriel amine synthesis** has been succesfully married with malonic ester chemistry in the synthesis of amino acids.

e.g.

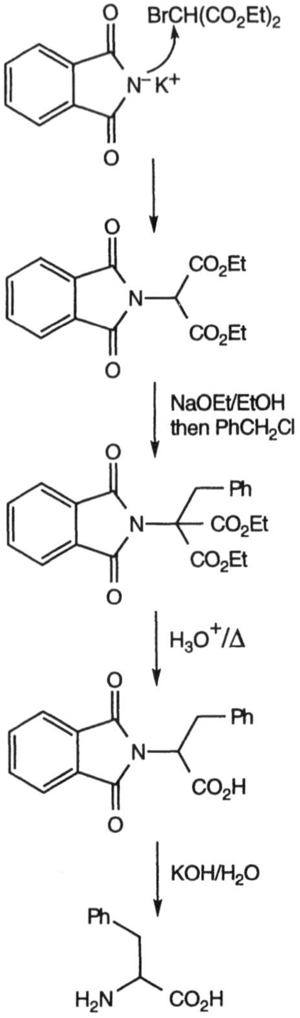

phenylalanine

conjugated with the carbonyl groups, and it doesn't have the option of reforming an anion...so it only acts as a nucleophile once. The full procedure is therefore:

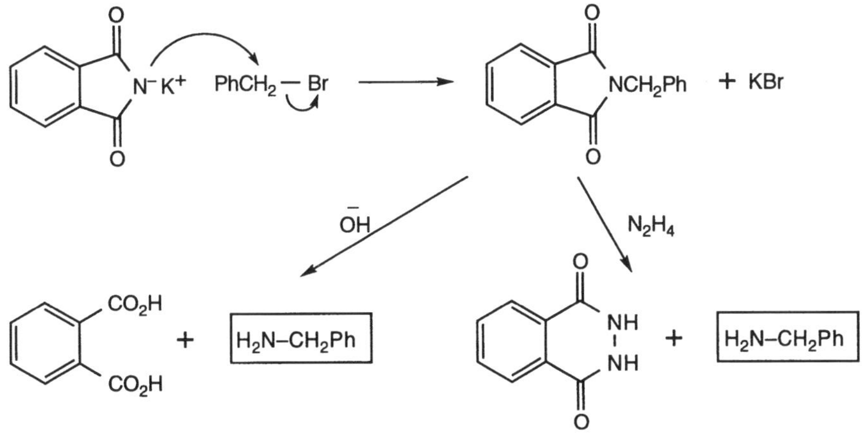

(b) CF₃CONH₂/NaH, then OH⁻ (aq)

This is the same type of reaction as the Gabriel procedure, but the final hydrolysis step is much milder. The sodium hydride deprotonates the trifluoroacetamide; the adjacent carbonyl group (−M) and nearby CF₃ group (−I) help to stabilize the charge, and ensure that the nitrogen acts as a nucleophile rather than as a base.

Trifluoroacetamide itself is both non-basic and non-nucleophilic. This is because the lone pair on the nitrogen is delocalized onto the carbonyl (see Chapter 7) and because the CF₃ group is so electron-withdrawing.

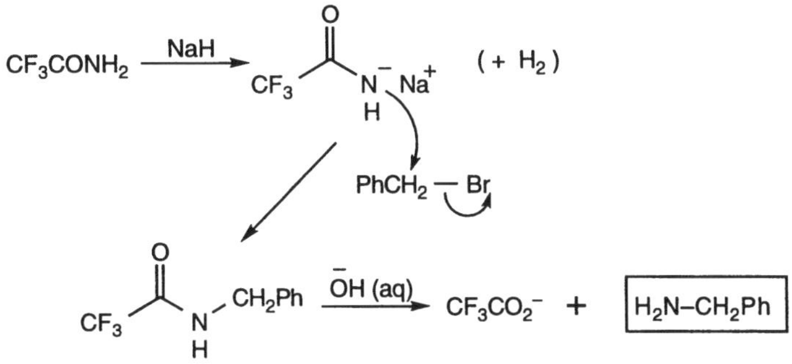

Once the trifluoroacetamide anion has acted as a nucleophile, its reactivity is lost (see Chapter 7 on amides), and it does not easily lose another proton. Amides are hydrolysed a bit more readily than imides (cf. Gabriel's reagent), but the electron-withdrawing CF₃ group increases that rate of hydrolysis further—so these species are about as reactive as esters. Thus, treatment with sodium hydroxide solution (1 M at room temperature) liberates the desired primary amine.

(c) NaN₃ then H₂/Pd−C

This is a justly popular method that usually proceeds in very good yield. Sodium azide is an excellent nucleophile, but will only react **once** (compare with amines). Reduction by catalytic hydrogenation then forms the free primary amine.

NaN₃ ⟶ [N̄=N⁺=N̄]   EtCH₂—Br ⟶ [EtCH₂–N=N⁺=N ... EtCH₂–N̄–N⁺≡N]

H₂N–CH₂Et + N₂ ⟵ H₂/Pd–C ⟵

The (perhaps surprising) stability of these dipolar azides is discussed in Chapter 10.

(d) (PhCH₂)₂NH then H₂/Pd–C

A neat variation is to use the chiral amine shown below (X), which undergoes conjugate addition to α,β-unsaturated esters with high asymmetric induction:

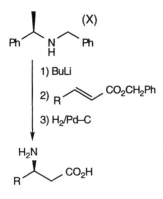

(From, M. E. Bunnage, A. N. Chernega, S. G. Davies, and C. J. Goodwin; *J. Chem. Soc., Perkin Trans. 1*, 1994, 2373.)

If secondary amines can be readily alkylated to tertiary amines, then a secondary amine with **removable** alkyl groups should be a good selective reagent for making primary amines. As the benzyl group can be removed from amines by hydrogenation, dibenzylamine can fulfil this role.

(PhCH₂)₂N̈H   Et—Br $\xrightarrow{\text{NaHCO}_3}$ (base to remove HBr)   (PhCH₂)₂N–Et

↓ H₂/Pd–C

2 x PhCH₃ + H₂N–Et

All of the above methods used nitrogen as a nucleophile in a key substitution reaction, although a subsequent step was then needed to free the amino group. One last substitution trick introduces the 'CH₂NH₂' unit.

(e) Use of KCN, followed by reduction

Cyanide is another good inorganic nucleophile that can only react once. This time, it is a nucleophilic carbon rather than nitrogen that displaces the leaving group, and the product therefore gains an extra carbon atom. The nitrogen can be liberated as the free NH₂ by reduction (lithium aluminium hydride (LiAlH₄) or H₂/Raney nickel under pressure), leading to the overall introduction of the CH₂NH₂ unit.

PhCH₂Br + KCN ⟶ PhCH₂CN + KBr

↓ LiAlH₄ or H₂/catalyst

PhCH₂CH₂NH₂

4.3 Amines via reduction reactions

There are a range of nitrogen functional groups that can be readily reduced to the corresponding amine. In this section, we won't discuss in detail how the nitrogen functional groups can be incorporated, or how they enable other parts of the structure to be built up—we will simply list some of those that can be reduced to the amine, and then discuss them briefly.

				$LiAlH_4$	$NaBH_4$	H_2/catalyst
(a) imines				✓	✓	✓
(b) amides				✓	X	X
(c) nitro groups				✓	X	✓
(d) nitriles				✓	X	✓

(a) Reduction of imines

This simple reduction can be achieved by catalytic hydrogenation (H_2/Pd–C) or by a hydride reducing agent—sodium borohydride ($NaBH_4$) is widely employed.

e.g.

The monobenzyl final product would have been very difficult to prepare using simple alkylation methods.

Importantly, it is often possible to form the imine *in situ*, and to carry out imine formation/reduction in a 'one-pot' reaction. In particular, this can be exploited when ammonia is the amine, allowing aldehydes and ketones to undergo overall reductive amination to the primary amine.

e.g.

The source of the ammonia 'NH_3' is usually aqueous ammonium acetate ($CH_3CO_2^-\ NH_4^+ \rightleftharpoons NH_3 + CH_3CO_2H$) which generates the imine in an initial equilibrium step. The sodium cyanoborohydride is a less reactive hydride source than NaBH$_4$, and is able to reduce the imine to the amine without reducing the ketone to the alcohol (see Chapter 5).

A variation to this approach is the Leuckart reaction, which effectively uses methanoic acid (formic acid) as the reducing agent:

The usual procedure is to employ the alkyl ammonium formate salt (RNH$_3$$^+$·HCO$_2$$^-$); this reacts with aldehydes and ketones as if it were a mixture of the amine and formic acid, thereby generating a new amine in a 'one-pot' reaction. An excess of the ammonium salt is required to avoid mixtures of amine products.

e.g.

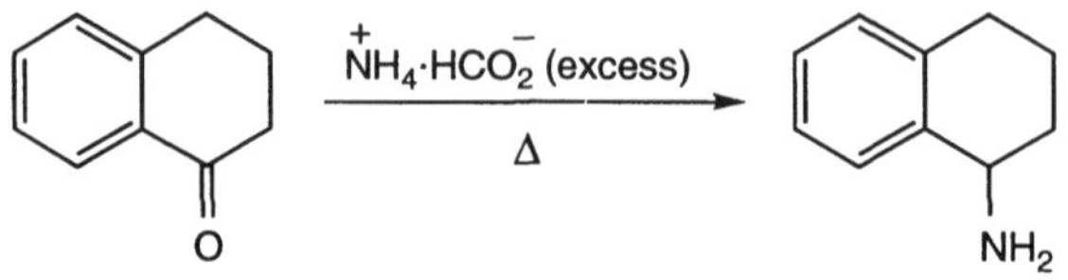

(b) Reduction of amides

Amides are readily formed, but are less readily reduced; LiAlH$_4$ is the usual reducing agent. Amides can be formed from amines by acylation with an acid chloride or acid anhydride (Chapter 7). The nitrogen in the product can delocalize its lone pair on to the carbonyl group; this stabilization explains why monoacylation is easily accomplished, but this stability is also the reason that such powerful reducing agents are required.

e.g.

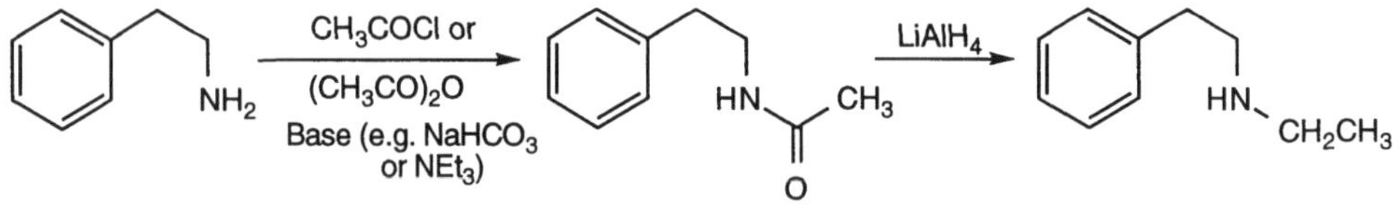

The base in the first step ensures that the acylation goes to completion. Once again, the overall process is monoalkylation of a primary amine—a transformation that is very hard to achieve by direct alkylation. However, because LiAlH$_4$ is such a powerful reducing agent, it is important to check that other unwanted reductions won't also occur in the final step.

(c) Reduction of nitro compounds

Nitro compounds are readily reduced by a huge range of reagents:

e.g.

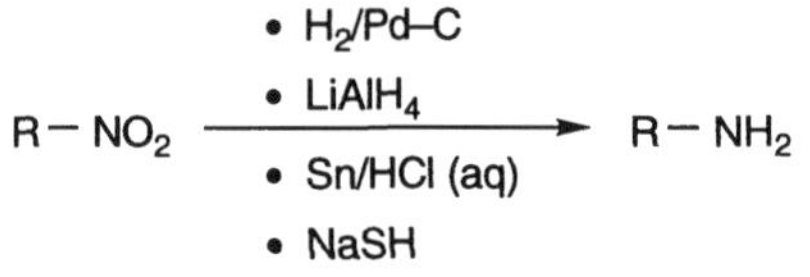

$$R-NO_2 \xrightarrow{\begin{array}{l}\bullet\ H_2/Pd-C \\ \bullet\ LiAlH_4 \\ \bullet\ Sn/HCl\ (aq) \\ \bullet\ NaSH\end{array}} R-NH_2$$

Aromatic nitro compounds are, of course, readily accessible by direct nitration of a benzene derivative, so this is a particularly attractive way of making aromatic amines.

e.g.

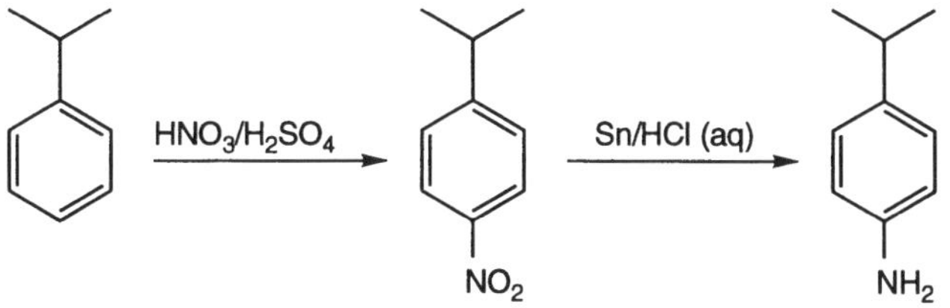

Although most of the reduction methods outlined above **do** work with aromatic nitrocompounds, the use of tin and hydrochloric acid is common, and this is the classical way of preparing aniline (phenylamine) from benzene. For aliphatic nitro compounds, the other reductive methods (especially catalytic hydrogenation) are used more often.

(d) Reduction of nitriles

This was mentioned when cyanide was discussed as a nucleophilic source of nitrogen. In fact, nitriles can be further elaborated before the $-C\equiv N$ group is reduced (see Chapter 8), so this is a particularly useful trick for preparing compounds containing the $-CH_2NH_2$ unit.

e.g.

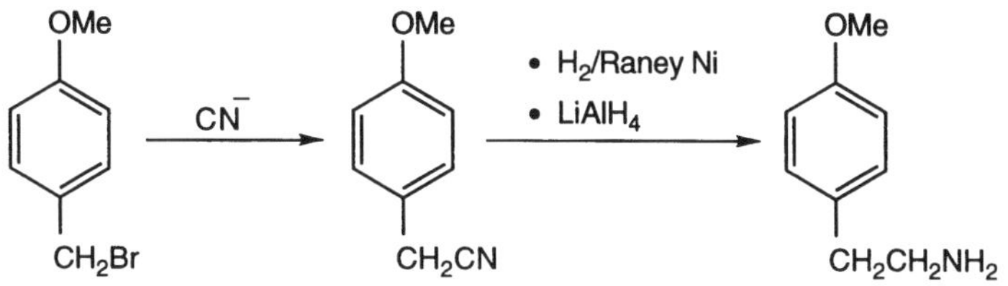

As the $-CN$ group is rather stable, powerful reduction methods are required such as $LiAlH_4$, or hydrogenation under pressure over Raney nickel (a highly activated nickel catalyst).

Complex amines from nitriles
As we will see in Chapter 8, nitriles can act like carbonyls and stabilize an adjacent negative charge:

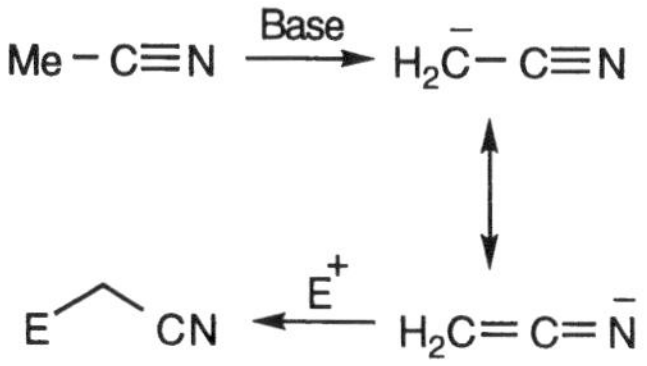

Thus the nitrile could be further functionalized prior to the reduction leading to relatively complex amine side-chains.

4.4 Summary

There are many methods of preparing amines, but there are essentially just two approaches (which are often combined):

- Nucleophilic displacement using a nucleophilic source of nitrogen (e.g. ammonia, 'NH$_3$' equivalent (Gabriel, NaN$_3$, etc.), RNH$_2$, CN$^-$).
- Reduction of a nitrogen-containing functional group (e.g. imine, amide, nitro, nitrile).

Unsaturated nitrogen compounds

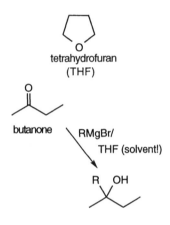
This section concerns unsaturated nitrogen compounds that do not possess N–N or N–O bonds. Although there are many subtle factors that affect the detailed reactions of organonitrogen compounds, the underlying chemistry of unsaturated nitrogen compounds is very similar to that of their oxygen counterparts. In particular, analogy with the chemistry of carbonyl compounds is a theme that will appear again and again. However, before looking at unsaturated nitrogen-containing groups, there are three general factors that we will consider.

(a) Polarization of C–N/C=N bonds

The chemistry of organonitrogen compounds is strongly influenced by the strength of the C–N single bond, and the reactivity of the C=N double bond. We can see the oxygen analogy with the isomeric ether (THF) and ketone (butanone) which are both C_4H_8O.

Whilst THF is commonly used as a solvent for organic reactions, butanone is a fairly reactive ketone that reacts (for example) with Grignard reagents. The difference in reactivity is caused by the C=O π-bond. Perhaps surprisingly, the π-bond is **stronger** than the σ-bond; this is because the polarizable π-bond can be distorted, so that greater electron density resides on the more electronegative oxygen atom. If the electron-deficient carbon atom is then attacked by a nucleophile, little additional energy is required to completely push the negative charge on to oxygen, and consequently the activation energy for the reaction is low. An extreme way of expressing this is to suggest that ketones possess a strong contribution from an ionic resonance form, for which nucleophilic attack on the carbon atom would be expected.

Typical bond strengths (kJ mol^{-1}):

C=O	730	C–O	360
C=N	615	C–N	305
C=C	610	C–C	345

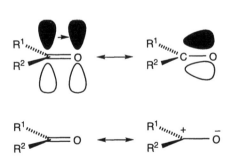

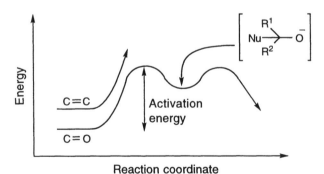

For C=N, the polarization is less pronounced, because the electronegativity difference between C and N is smaller than that between C and O. Nucleophilic attack on an imine would generate an intermediate with negative charge on nitrogen, and this is less favourable than negative charge on oxygen, so imines are somewhat less reactive than ketones. However, imines can be readily **activated** by the addition of catalytic acid. Protonation of the imine gives an **iminium** intermediate. These species are activated because the C=N π-bond is more polarized and because nucleophilic attack now generates a neutral intermediate:

Strong acids can effectively fully protonate imines at pH 1, whilst less than 1% of a carbonyl compound is protonated at this pH. Nevertheless, nucleophilic attack on carbonyl compounds can still be acid catalysed, as the small equilibrium quantity that is protonated undergoes reaction.

iminium ion

So the C=N π-bond is less stable than the C=O π-bond, and protonation of C=N activates it towards nucleophilic attack.

(b) Enol/keto equilibria

The chemistry of carbonyl compounds is dominated by their ability to exist as enol or keto tautomers, and the ease with which they can be deprotonated to give resonance stabilized enolates:

The enol/keto equilibrium usually lies predominantly as C=O, because of the stability of the carbonyl π-bond. However, the enol is nucleophilic, and electrophiles can trap out the enol by reaction at oxygen or carbon, although most reactions occur on carbon because the C=O π-bond is thereby regenerated:

Powerful electrophiles are needed, as enols are only weak nucleophiles. In contrast, a deprotonated enol (i.e. an enolate) is negatively charged, and is a much more powerful nucleophile—either on carbon or (less frequently) on oxygen:

Hard and soft nucleophiles / bases
When a reactant uses its lone pair to bond to a proton in a co-reactant it is acting as a **base**. When it attacks any other electron-deficient atom (usually carbon), it is said to act as a **nucleophile**. In general, highly charged ('hard') nucleophiles react fastest with highly charged electrophiles whilst polarizable ('soft') nucleophiles attack polarizable electrophiles fastest. Moreover, as the C–H bond is quite strongly polarized (and H is small), hard nucleophiles often react preferentially as bases. This is really just a fuzzy way of summarizing the molecular orbital factors that control these reactions (see I. Fleming, *Frontier orbitals and organic chemical reactions*, Wiley, 1976).
Thus, many anions that have localized charge act preferentially as bases, whereas neutral or delocalized anions act as nucleophiles.
e.g.

$$CH_3CH_2Br$$

HNEt₂ / NaNEt₂

$$CH_3CH_2NEt_2 \qquad H_2C=CH_2$$

Enolates can act as nucleophiles on carbon or oxygen. The ratio of C/O attack depends on matching the nucleophilic/electrophilic characters, and can be controlled by choice of electrophile, enolate counter-ion, and solvent.

The reactions of C=N are very similar. Thus:

imine enamine iminium

We will discuss these reactions at greater length in Chapters 5 and 6. But it should be clear that enamines are the nitrogen equivalents of enols: they are more reactive than enols (nitrogen is a better nucleophile, being more willing to accept a positive charge than oxygen), but less reactive than enolates, which carry a negative charge. This intermediate reactivity can be profitably exploited in the synthetic reactions of enamines.

Double bond equivalents (DBE)
Related to oxidation level is the concept of double bond equivalents. Each π-**bond** and each **ring** counts as a DBE. Reduction usually leads to a decrease in DBE, whilst oxidation often leads to an increase.
To calculate DBE in a covalent structure:
Count the rings and π-bonds:
e.g.

$C_{11}H_7N_2OCl$ DBE = 9

To calculate DBE from a molecular formula:
(1) Formula of saturated parent is C_nH_{2n+2}; use number of carbons to calculate $(2n+2)$.
(2) Count hydrogens in actual formula and amend them as follows:
• For each valency-1 atom (e.g. halogens), add 1.
• For each valency-2 atom (e.g. oxygen), do nothing.
• For each valency-3 atom (e.g. neutral N), subtract 1.
• For each valency-4 atom (e.g. Si, N^+), subtract 2.
  etc.
This gives C_nH_x (x = amended H count)
(3) Finally, **DBE = $[(2n+2)-x]/2$**
e.g. $C_{11}H_7N_2OCl$
Parent hydrocarbon : $C_{11}H_{24}$
Amended $C_{11}H_7N_2OCl$ gives $C_{11}H_6$
so DBE = $(24-6)/2 = 9$.

(c) Oxidation levels

When trying to work out a suitable reagent for achieving a particular transformation, it is useful to be able to determine rapidly whether the required process involves oxidation, reduction, or neither. The oxidation level of a functionalized carbon atom is simply the number of heteroatom bonds attached to that carbon atom. Thus, in $R-C\equiv N$ the nitrile carbon has three bonds to nitrogen, and therefore has an oxidation level of 3. Since carboxylic acids also have an oxidation state of 3, the interconversion of nitriles and acids does not require oxidation or reduction.

e.g.

$RCN \longrightarrow RCO_2H$
No change in oxidation level
Reagent: acid or base hydrolysis

$RCONH_2 \longrightarrow RCN$
No change in oxidation level
Reagent: powerful dehydrating
agent (PCl_5)

$RCONH_2 \longrightarrow RCH_2NH_2$
Oxidation level 3 $\longrightarrow$ 1
Reagent: $LiAlH_4$

The following table summarizes the oxidation state of carbon in a range of nitrogen and oxygen functional groups:

Oxidation level	Nitrogen-free functional groups	Nitrogen functional groups
0	C—C—C with C above and below central C (quaternary carbon)	
1	\CHOH \CHCl \CHBr \CHI	\CHNH$_2$
2	\C=O ⇌ (\C—OH) \C(Br)(Br)	—C(H)=NR ⇌ (\C(H)—NR(H)) —C(OR')(H)—NHR
3	\C(RO)=O \C(Cl)=O	—C≡N \C(RN)(=O) (cyclic) \C(R'O)=NR \C(RN,H)=O
4	CO$_2$ \C(RO)(RO)=O	O=C(RO)(NR'H) O=C(RNH)(NR'H) RN=C=NR'

Notice that enols and enamines are given an oxidation state of 2. This is because the enol/keto-type equilibrium can be exploited in their synthesis or reactions. In all other cases, the simple oxidation state analysis works.

Summary

We are now in a good position to look at the chemistry of a range of unsaturated nitrogen functional groups, bearing in mind:

- Comparison with oxygen functional groups.
- Information on C–N/C=N bond polarization.
- · Simple rules on oxidation level.

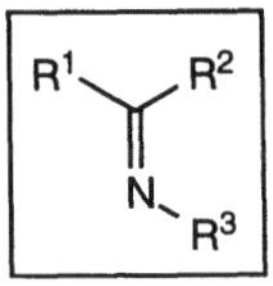

5 Imines

5.1 Key features of imines

- Oxidation level 2.
- Susceptible to nucleophilic attack on carbon.
- Imines are easily protonated to produce reactive iminium ions.

5.2 Synthesis of imines

Imines are simply prepared from the reaction of an aldehyde or a ketone with a primary amine.

e.g.

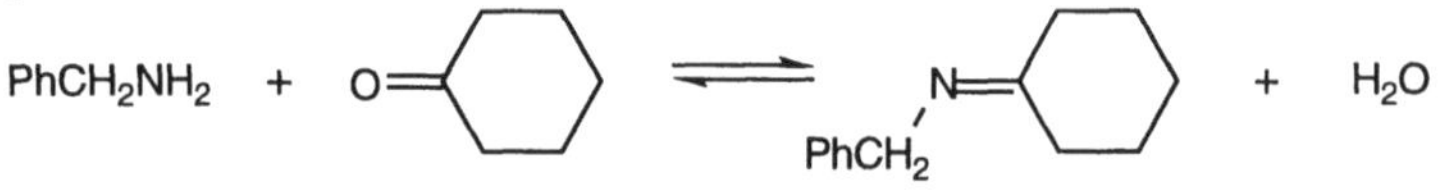

As the C=N π-bond is rather weak, the above equilibrium lies to the left if there is excess water present. In order to obtain a good yield of imine, the water generated in the reaction must be specifically removed. The use of molecular sieves or azeotropic removal of water using a Dean–Stark apparatus are common methods.

If the imine is conjugated to an aromatic ring, then it is more stable than simple aliphatic imines. However, most imines are used immediately because of their limited stability on storage.

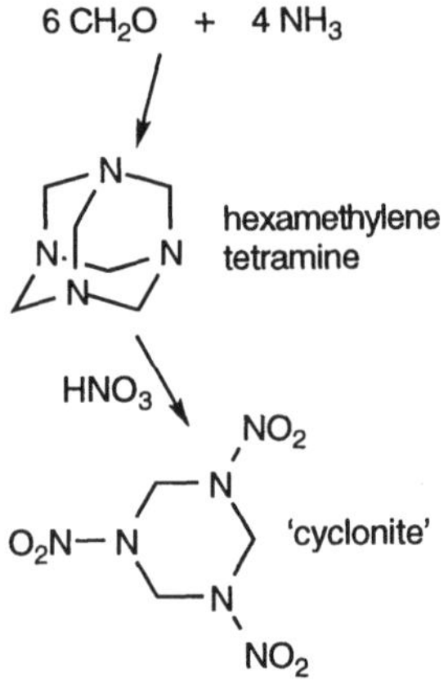

5.3 Reactions of imines

Attack by carbon nucleophiles

Stable imines can be treated with a range of carbon nucleophiles, and this is a useful method for synthesizing amines:

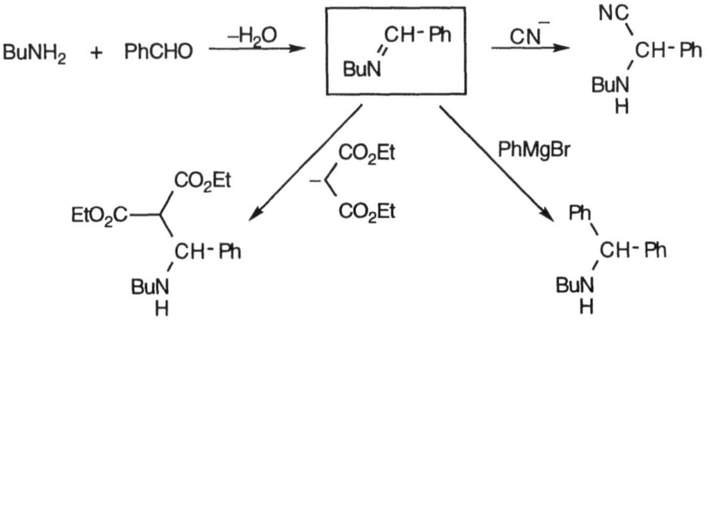

Storing imines

One problem with trying to store imines is polymerization—the equilibrium with enamines (see Chapter 6) allows aldol-type condensations to occur. Of course, related reactions have been exploited, for example in the formation of hexamethylene tetramine. This solid was the first organic substance whose structure was determined by X-ray crystallography. Treatment with nitric acid generates 'cyclonite' which was widely used as a high-explosive in the Second World War.

$$6\ CH_2O\ +\ 4\ NH_3$$

hexamethylene tetramine

For most nucleophiles, it isn't possible to trap the imine *in situ*, because other reactions can take place with the amine or carbonyl starting materials. However, it is possible to trap the imines *in situ* in several important variants of the **Mannich** reaction (see a and b below) and in the **Strecker** reaction (see margin).

(a) *Intermolecular trapping of iminium ions.* In general, imines that are substituted on carbon are too hindered and unreactive to be trapped *in situ*. But the iminium ions derived from methanal are completely unhindered, and they can be trapped by many neutral carbon nucleophiles.

e.g.

These reactions are successful not only because of the unhindered nature of the iminium ion, but also because it holds a full positive charge and is thus very reactive. Notice that enols are sufficiently nucleophilic—the overall process is then effectively a crossed-aldol condensation. Iminium ions are also trapped in reductive amination reactions, such as $Na(BH_3CN)/NH_4OAc$ or Leuckart reactions (see Section 4.3). With primary amines, the neutral imine is formed, which is less reactive.

(b) *Intramolecular trapping of iminium ions.* When a carbon nucleophile is **attached** to an imine, the Mannich reaction can take place very readily. This process is one of the most important reactions for the biosynthesis of alkaloids, which are complex naturally occurring nitrogen compounds. Where a five- or six-membered ring is being formed, imines are usually trapped very efficiently, provided catalytic acid is present to activate the imine as the iminium derivative.

This reaction, with an unactivated aromatic ring, is rather sluggish. With phenolic or indolic aromatic compounds, the cyclization is fast and efficient. This is the case in the synthesis and biosynthesis of many alkaloids, and some routes to isoquinoline and yohimbine alkaloids are outlined as follows:

Mannich reactions involve the trapping of an iminium ion by a carbon nucleophile.

The **Strecker synthesis** of amino acids **does** involve *in situ* trapping of an imine.
e.g.

$PhCHO + NH_3 + HCN$

phenylglycine

This methodology is described in more detail in Chapter 8.

Iminium electrophiles
Although the chemistry is effectively that of imines, iminium species (e.g. A or B) are much more reactive.

$R_2NH + R'_2CO$

Can rearrange to an enamine if R' has an α-proton (see Chapter 6).

A

$RNH_2 + R'_2CO$

B

For a review of alkaloid synthesis using related cyclizations, see H. Hiemstra and W. N. Speckamp, *The Alkaloids*, 1988, **32**, 271.

R = H, Me; isoquinoline
alkaloids

indole alkaloids
e.g. yohimbine

Similarly, intramolecular trapping of imines by enols is usually efficient, again leading to alkaloid substructures.

e.g.

(From W. G. Early, T. Oh, and L. E. Overman, *Tetrahedron Lett.,* 1988, **29**, 3785.)

TFA

The Vilsmeyer reaction

This reaction is a simple extension of the basic Mannich reaction, in which dimethylformamide (DMF) is transformed into an iminium intermediate by reaction with $POCl_3$:

Attack be Cl⁻ may be intermolecular(as shown) or intramolecular (cf. $ROH + SOCl_2$).

The iminium ion can be trapped by carbon nucleophiles, but the work-up procedure leads to loss of the nitrogen and chlorine, and formation of an aldehyde:

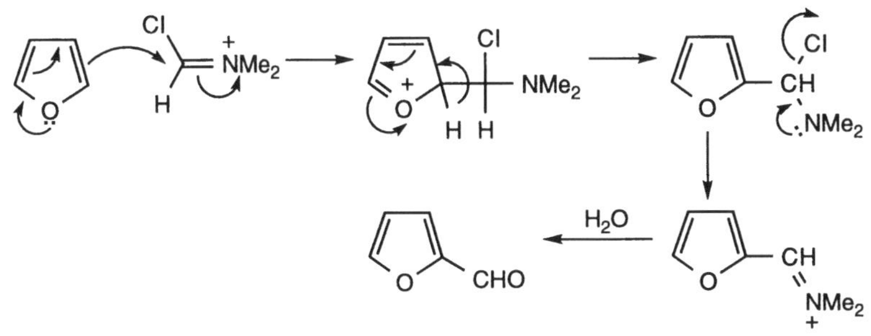

Strictly speaking, this is a slightly different class of reaction—the extra chlorine increases the oxidation level of the reactive iminium species to 3, so that it is at the 2 oxidation level at the end of the reaction. The final oxidation level of 2 is the same as aldehydes and explains why, after hydrolysis, the CHO group is introduced overall.

Reduction of imines

Imines can be reduced to amines using a wide variety of reducing agents:

$$R-CH=NR' \xrightarrow{[H]} R-CH_2-NHR'$$

The most commonly used reducing agents are catalytic hydrogenation (H_2/Pd–C) or sodium borohydride—one of the mildest hydride reducing agents. Reduction of imines is one of the best methods of effecting overall monoalkylation of a primary amine.

e.g.

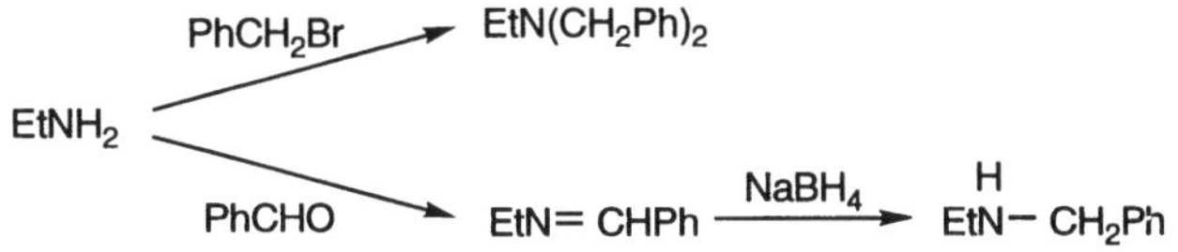

In this case hydrogenation would not be appropriate since we would cleave the benzyl group and merely regenerate starting material!

Using this reduction trick we can also introduce a primary amino group:

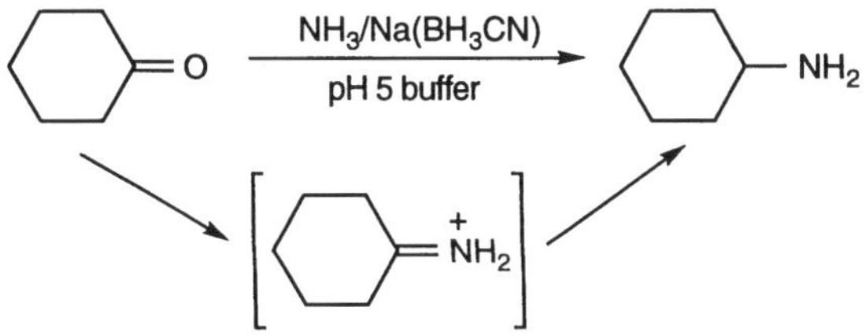

The conditions for this reaction are critical, but it is a pretty reliable method of converting aldehydes and ketones into the corresponding amines. The cyanide in sodium cyanoborohydride lowers the activity of the reagent so that aldehydes and ketones are not reduced. Imines would also be unaffected, except the pH is just low enough for some protonation to the iminium ion to occur,

Reduction of C=C and C=O

(a) The non-polarized C=C π-bond is reduced by unpolarized H–H and a catalyst; C=O bonds are usually stable to such conditions.

(b) Polarized C=O π-bonds are reduced by nucleophilic sources of H^- (e.g. $LiAlH_4$, $NaBH_4$); C=C π-bonds are usually stable.

(c) C=N π-bonds are less polarized than C=O bonds, but more polarized than C=C bonds. Both types of reduction usually work.

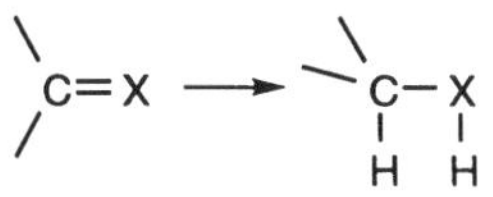

	H_2/Pd–C	$NaBH_4$
$X = CR_2$	✓	✗
$X = O$	✗	✓
$X = NR$	✓	✓

Diels–Alder reactions have not been included under the main coverage of 'reactions'. The π-bond can be used in (4+2) cycloadditions, but conditions and reactants must be carefully chosen. For example:

(major product)

(From P. D. Bailey, R. D. Wilson, and G. R. Brown, *J. Chem. Soc., Perkin Trans. 1*, 1991, 1337.)

In a similar way, β-**lactams** are accessible by (2+2) cycloaddition of the imine π-bond to a ketene (often generated *in situ*). For example:

(From S. S. Bari, I. R. Trehan, A. K. Sharma, and M. S. Manhas, *Synthesis*, 1992, 439.)

and reduction of this species is rapid. The solution is usually buffered with ammonium acetate—and this also cleverly provides the source of ammonia.

Rather like the intramolecular Mannich reaction, imines and iminium ions within a cyclic system are particularly easy to form and trap out, as illustrated here:

Hydrolysis of imines

Remember that imines and iminium ions are readily formed from amines and aldehydes or ketones, but this reaction is reversible. Indeed, the equilibrium lies over towards hydrolysis unless water is explicitly removed. Although this is sometimes a nuisance, you will see that the chemistry of imines can sometimes be exploited, with the intention of carrying out a hydrolysis at the end of the reaction. This is exemplified by the use of enamines, which are discussed in the next chapter.

5.4 Summary

- Imines are formed (reversibly) by reaction of a primary amine with an aldehyde or ketone, and the reaction is driven by removal of water.

- Imines react with carbon nucleophiles to form amines, or can be trapped out in the Mannich or Strecker reactions:

- Reduction of imines to amines is readily achieved using hydrogenation or hydride reducing conditions:

6 Enamines

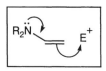

When primary amines are reacted with aldehydes or ketones, we have seen that an imine is formed provided water is removed. This reaction proceeds via an iminium intermediate (X).

e.g.

If a secondary amine is used instead, the same type of intermediate (Y) can be formed, but now there is no proton on the nitrogen that can be lost. The only way that neutrality can be regained is to lose a proton **adjacent** to the iminium moiety.

e.g.

The product from this reaction is an enamine. Perhaps surprisingly, these compounds rarely act as nucleophiles (or bases) on nitrogen. Presumably the following resonance hybrids render the nitrogen $\delta+$ and the β-carbon $\delta-$.

Instead, enamines almost always react as nucleophiles at the β-carbon, and this property has rendered them extremely valuable in synthesis.

You will probably have realized that enamines are the nitrogen analogues of enols. They are more reactive than enols because the nitrogen is better at holding a positive charge, but are less reactive than negatively charged enolates. This intermediate reactivity makes them readily prepared, but quite powerful reagents for synthesis.

Increasing nucleophilicity

Using enamines in synthesis requires three distinct stages: formation, reaction, and hydrolysis.

6.1 Formation of enamines

Enamine formation from an amine and an aldehyde or a ketone is actually quite slow. Their formation is acid catalysed (as with imines), and requires the actual removal of water—azeotroping in benzene or toluene is commonly used. In order to increase the rate of reaction, cyclic secondary amines are commonly employed, because the steric bulk of the amine is thereby reduced. So a typical procedure for forming an enamine might be:

Three widely used cyclic secondary amines are:

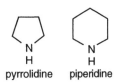

pyrrolidine   piperidine

morpholine

Aldehydes and symmetrical ketones can only form one regioisomer, but unsymmetrical ketones can form two possible regioisomers. In practice, the more substituted double bond is strongly favoured.

e.g.

major product

In general, enamines are formed under reversible conditions (thermodynamic control), and the most stable alkene results. So if (*E*)- or (*Z*)-isomers can be formed, it is the least sterically crowded isomer that predominates.

e.g.

major product

With the enamine in hand, we can use it as a nucleophilic source of carbon.

6.2 Reaction of enamines with electrophiles

Enamines are only moderately nucleophilic, so the electrophiles need to be quite reactive. The following reactions work well:

• *With alkyl halides* This is successful because alkyl halides can be heated quite strongly without serious side-reactions occurring. Provided the enamine possesses a β-substituent, monoalkylation is readily achieved.

e.g.

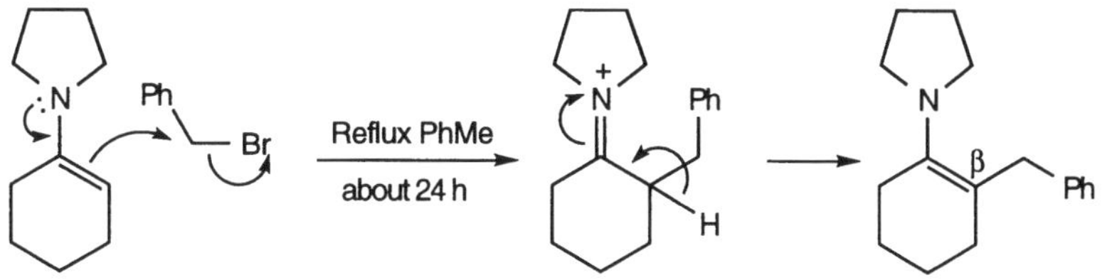

Although a nucleophilic enamine is regenerated, the β-carbon is now fully substituted, and steric factors preclude further reaction. This general principle applies to other reactions of enamines—in the following examples, the approximate reaction times indicate the relative reactivity.

• *With aldehydes*

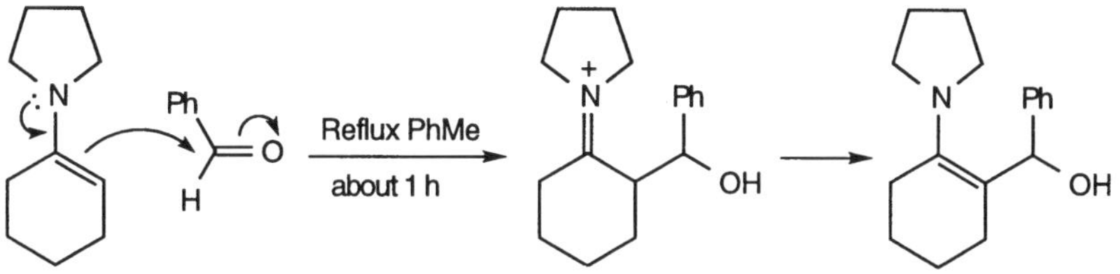

• *With ketones*

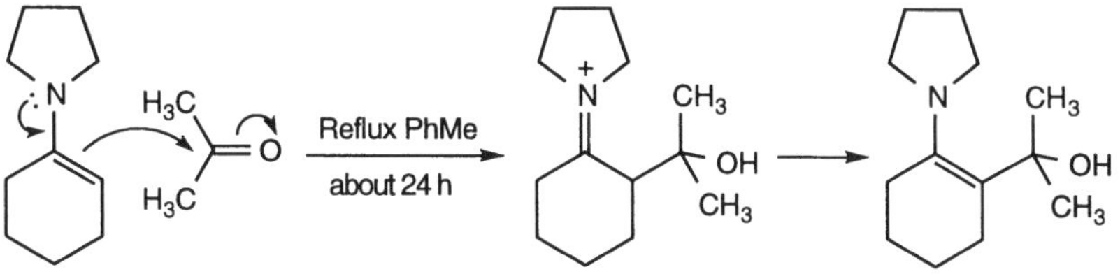

• *With α,β-unsaturated carbonyl compounds*

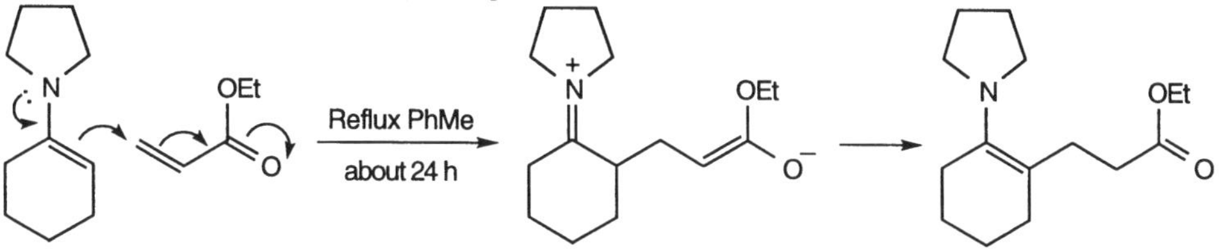

• *With acid chlorides*

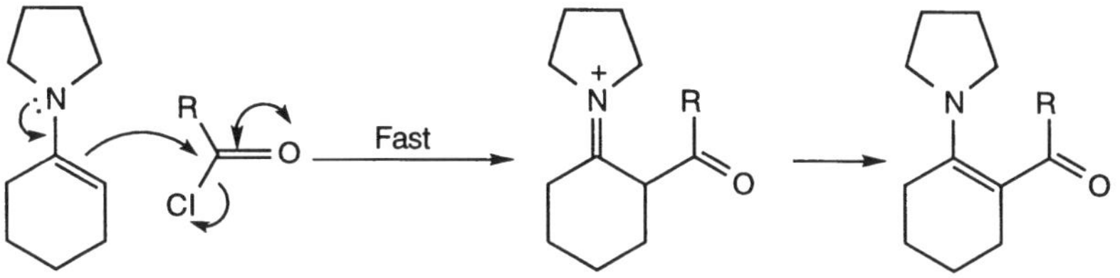

6.3 Hydrolysis of enamines

The product of quenching an enamine with an electrophile is...another enamine. Whilst the β-carbon may be too crowded for another carbon electrophile to approach, the much smaller proton (H⁺) can get in! If water is also present, then the resulting iminium ion is hydrolysed. So we can regenerate the original secondary amine and simultaneously unmask the carbonyl group. This reaction is simple and general, as summarized below:

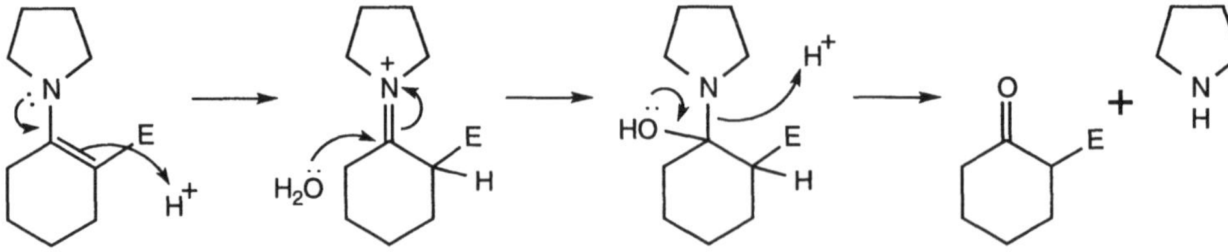

These work-up conditions do sometimes lead to further modifications of the product. For example, α,β-unsaturated carbonyl compounds can be formed by dehydration:

e.g.

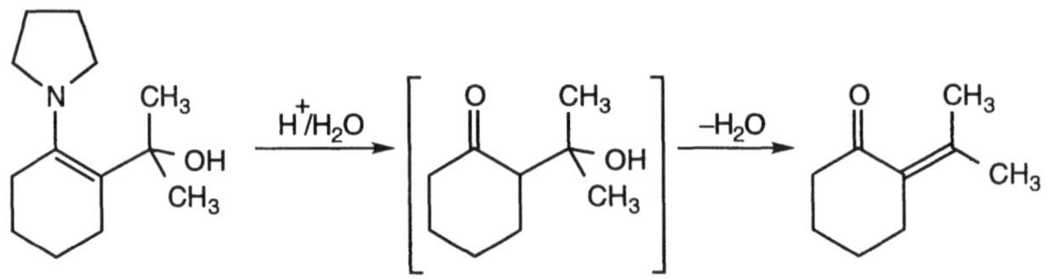

Typical enol/enolate chemistry

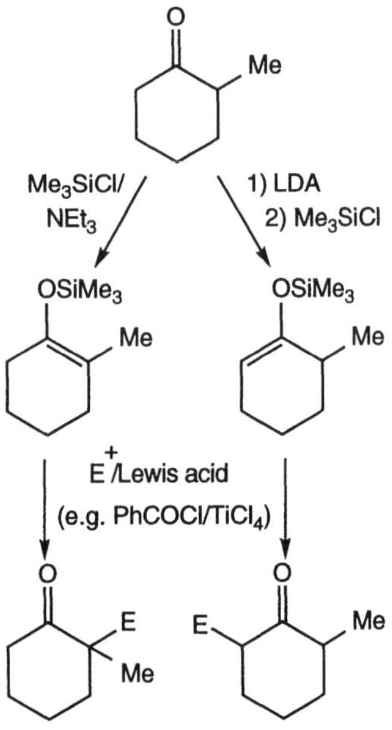

Although enamine chemistry was very attractive when first developed, enol/enolate chemistry has since advanced very rapidly and is now the more common method for placing electrophilic substituents adjacent to carbonyl groups.

But enamine chemistry **is** very widely used as a mild way of initiating intramolecular reactions:

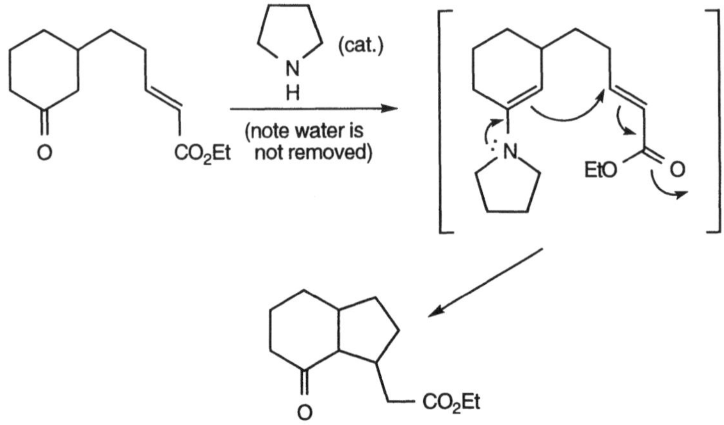

One particularly striking example of this is the formation of the decalin (X) by a cyclization that is triggered by the amino acid (L)-proline:

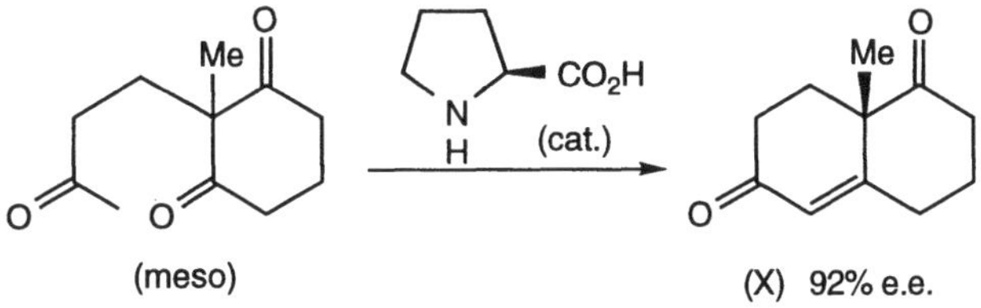

(meso)   (X) 92% e.e.

Of the various enamines that can form, only one is set up to cyclize readily, and the reaction rapidly proceeds down this pathway. The starting material is optically inactive (meso) but, amazingly, (L)-proline triggers asymmetric cyclization. The product contains about 96% of one enantiomer and only 4% of the mirror image (an enantiomeric excess of 92%). Unfortunately, catalytic asymmetric induction of such high efficiency is rare!

6.4 Enamine/imine tautomerism

Enamine/imine equilibria can be established in the same way as keto/enol equilibria:

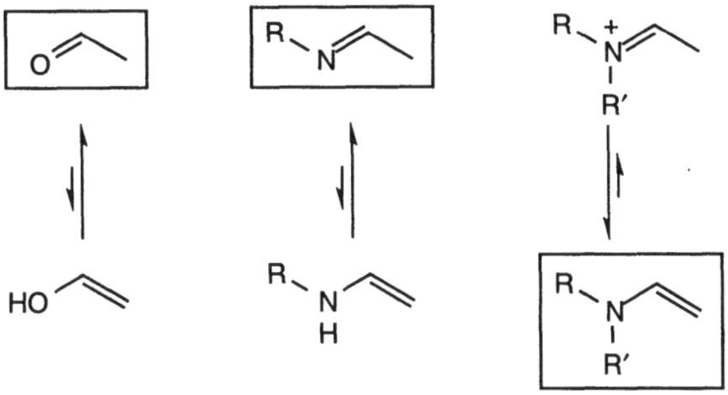

Importantly, primary amines react with aldehydes and ketones to give predominantly imines (heteroatom π-bond favoured). In contrast, secondary amines form enamines in order to give a neutral product. However, the fact that both processes are equilibria can be exploited, as outlined below.

Using imines as enamines

Primary amines **do** sometimes catalyse enol-like chemistry. This must be via imine/enamine tautomerism and it is essential that water is **not** removed. An efficient asymmetric example is:

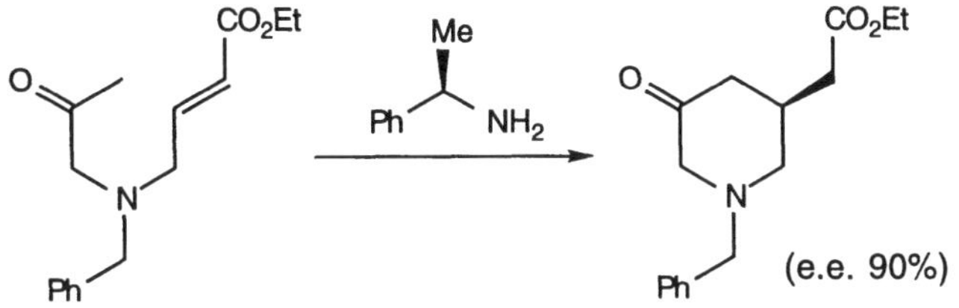

(e.e. 90%)

(From Y. Hirai, T. Terada, T. Yamazaki, and T. Momose, *J. Chem. Soc., Perkin Trans. 1*, 1992, 509.)

Q. What is the mechanism?

R. C. F. Jones *et al.* reported this synthesis of (*R*)-2-methylpiperidine in *Tetrahedron Lett.*, 1993, **34**, 6329.
Note:

① The enamine acts as a C-nucleophile onto the Michael acceptor.

② Reductions galore! Protonation of the enamine gives the imine which is reduced to an amine. This then condenses onto the carbonyl to make another imine, which is reduced *in situ*.
Q. After hydrolysis of the ester to the carboxylic acid, see how decarboxylation could occur...

③ The 1,1-diamine is in equilibrium with the corresponding amine and imine, which is reduced by 'H⁻':

④ Cleavage of the N-benzylic groups by catalytic hydrogenation.

Most piperidines have potent biological properties. For example, (*S*)-2-propylpiperidine is called coniine; it is the main alkaloid constituent of hemlock, which Socrates drank to commit suicide.

Using enamines as imines

So far, all the routes we have discussed have discarded the enamine nitrogen at the end of the synthesis. But in this neat example of rather complex enamines, the interplay between enamine and imine (actually iminium) is fully exploited in the asymmetric synthesis of N heterocycles.

6.5 Summary

The key processes associated with enamines are formation, electrophilic attack, and subsequent hydrolysis, as summarized below:

7 Amides

The amide bond links together the amino acids that make up all peptides and proteins. Not only is it a very robust functional group, but it has a range of intrinsic properties that help control the three-dimensional shape of peptides and proteins. We will therefore consider separately the properties, synthesis, and reactions of amides.

7.1 Features of amide bonds

- Planar
- Relatively inert
- Non-basic
- Weakly nucleophilic on oxygen
- Hydrogen bond acceptor and donor

All these features rely on the simple conjugation properties of amides:

For the overlap of N lone pair and C=O π-bond, they must be in the same plane

Planarity of amide bonds

Planarity is required for good orbital overlap. For most amide/peptide bonds, the bulky groups are *trans*. But secondary amines can form amides in which two isomers (called rotamers) can be seen by NMR, even though they interconvert hundreds of times per second at room temperature.

e.g.

dimethylformamide (DMF)

^{1}H NMR in CDCl$_3$ at RT
δ 2.7 (3H, s, CH$_3^A$)
δ 2.9 (3H, s, CH$_3^B$)
δ 8.05 (1H, s, CHO)

Even low molecular weight peptides can have interesting properties. For instance, L-aspartyl-L-phenylalanine methyl ester is better known as the sweetener Aspartame (A):

A slightly larger example is Met-enkephalin (B), a natural painkiller:

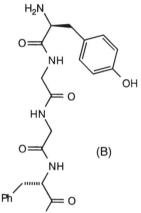

Of course amides are also present in many important natural products such as penicillin G (C):

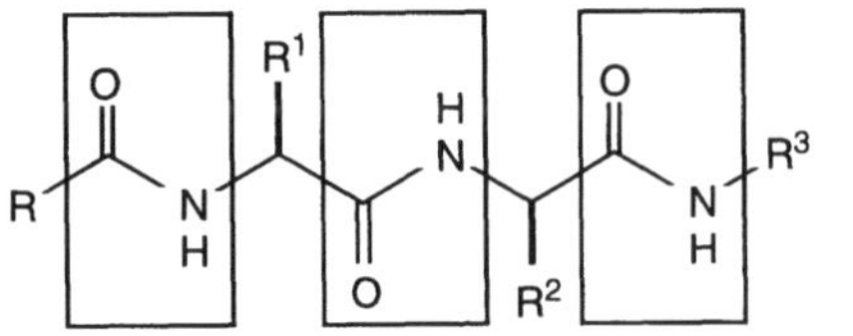

Planar peptide bonds

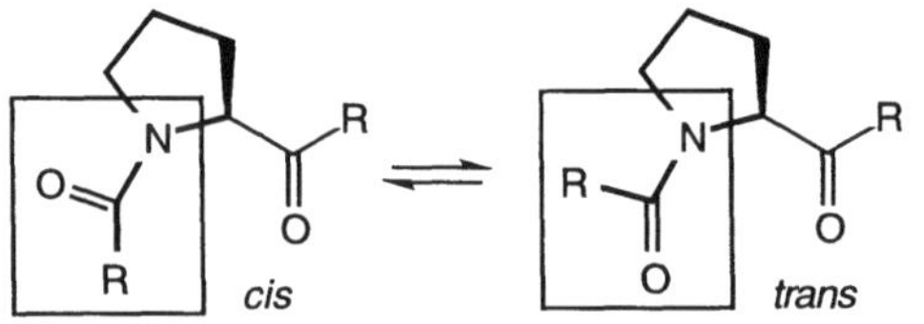

In contrast to normal *trans* peptide bonds, the proline residue shown here can be *cis* or *trans*

Inertness of amides

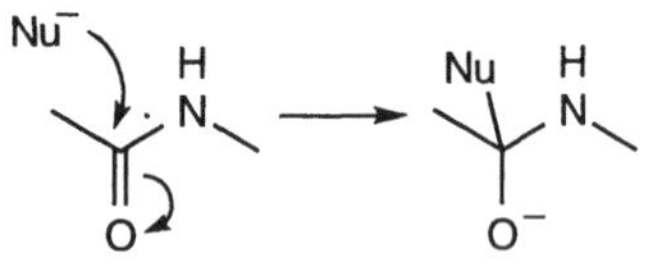

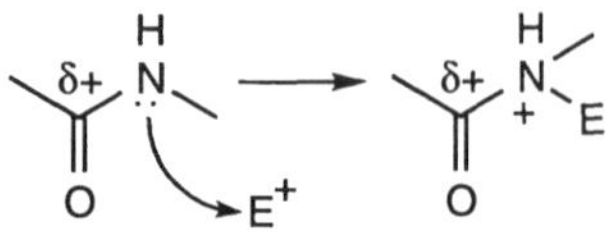

We might expect the amide nitrogen to be nucleophilic (or basic), and the carbonyl carbon to be electrophilic. But both these reactions have to pay the price of destroying the resonance stabilization and therefore occur only sluggishly. This inertness is often exploited—amines can be protected as amides or more often as urethanes (RHN–C(O)OR′, see Chapter 9).

The following reactions would cause loss of conjugation:

Weak nucleophilicity of amide oxygen

The resonance makes O δ– and N δ+. However, amides are very poor nucleophiles because such reactions involve loss of conjugation. Nevertheless, amides are weakly nucleophilic on oxygen, as seen in a limited number of intramolecular reactions (e.g. see page 51) and in the Vilsmeyer reaction.

Hydrogen bonding of amides

Resonance induced polarization makes the H-bonding of amides stronger than that of ketones or amines. In fact, a peptide chain is often locked into a conformation which maximizes the H-bonding, giving rise to secondary structure, such as β-sheets; which are key features of protein 3D structures.

Sulfonamides are related to amides:

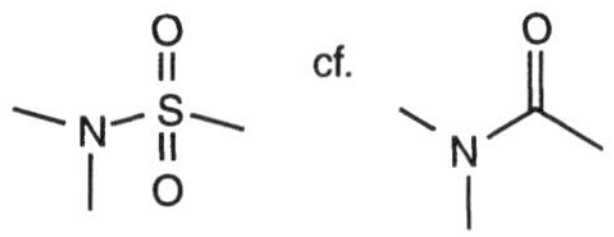

Examples include saccharin (D) and sulfonamide antibiotics (E).

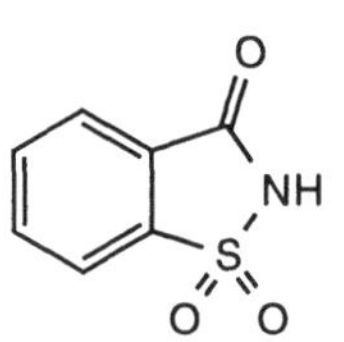

(D)

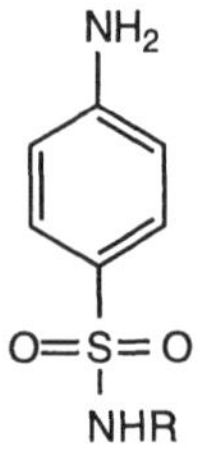

(E)

Anti-parallel β-sheet held together by H-bonding

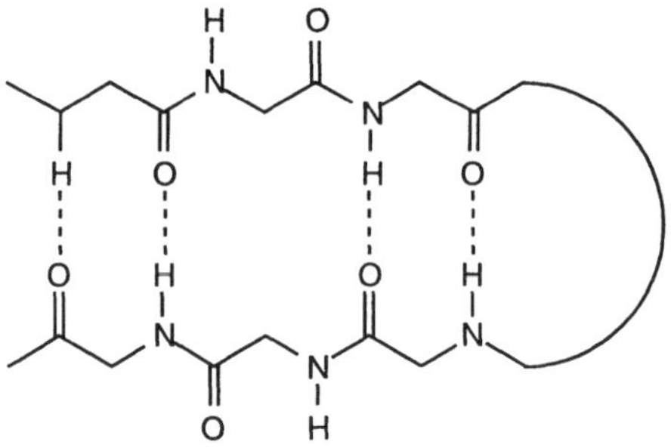

7.2 Synthesis of amides

Although there are many ways of making amides, one general procedure is dominant—the reaction between an amine and an activated carboxylic acid derivative:

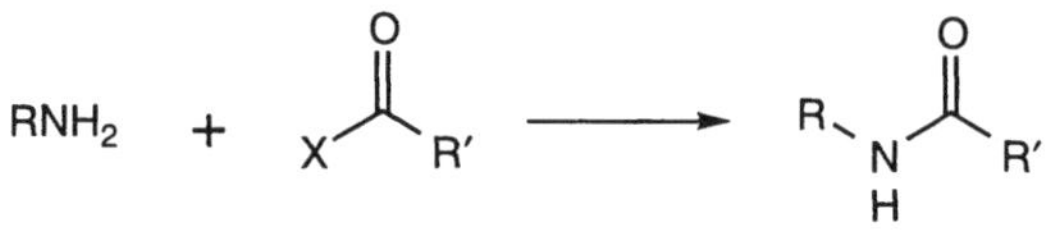

In the literature there are hundreds of possibilities for the nature of X, but three favourites are acid chlorides, acid anhydrides, and active esters:

X	Name	Leaving group
Cl	Acid chloride	Cl^-
OCOR″	Acid anhydride	$R''CO_2^-$
OAr	Active ester	ArO^-

The method you choose depends very much on the sort of amide that you are making.

Amides via acid chlorides

Acid chlorides are very reactive and are easily produced from a carboxylic acid by treatment with thionyl chloride ($SOCl_2$) or oxalyl chloride (($COCl)_2$). However, their extreme reactivity often causes problems, so they are rarely used except for forming simple acetyl amide derivatives (i.e. using CH_3COCl).

Amides via acid anhydrides

Acetic anhydride (($CH_3CO)_2O$ or Ac_2O) is widely used for forming simple acetyl amides. Other symmetrical anhydrides are easily prepared and are often used in peptide synthesis, because the amide bond formation is so clean and efficient:

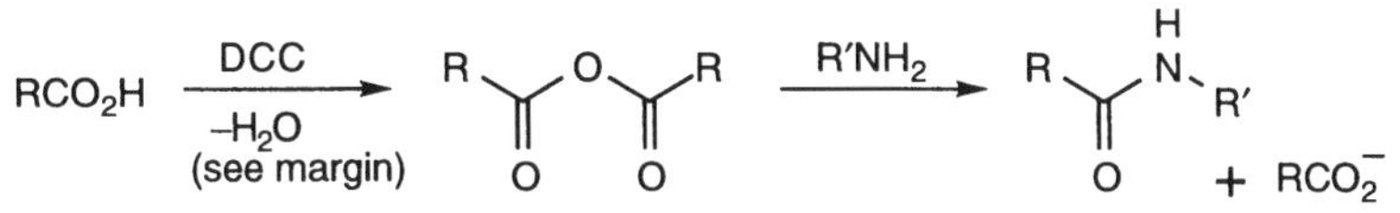

Dicyclohexylcarbodiimide (DCC) effectively removes one mole of water from two equivalents of the acid:

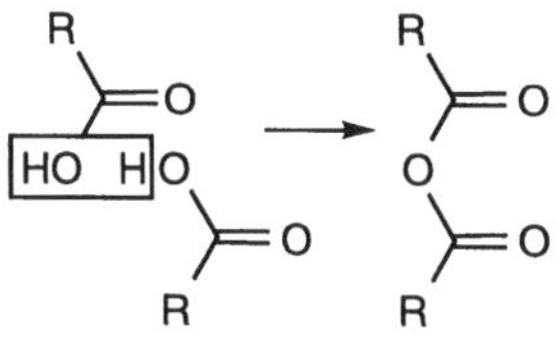

More details on the mechanism by which DCC effects dehydration are given in Chapter 9.

For unsymmetrical anhydrides there are two carbonyls that can be attacked—but in some cases one carbonyl is more reactive than the other:

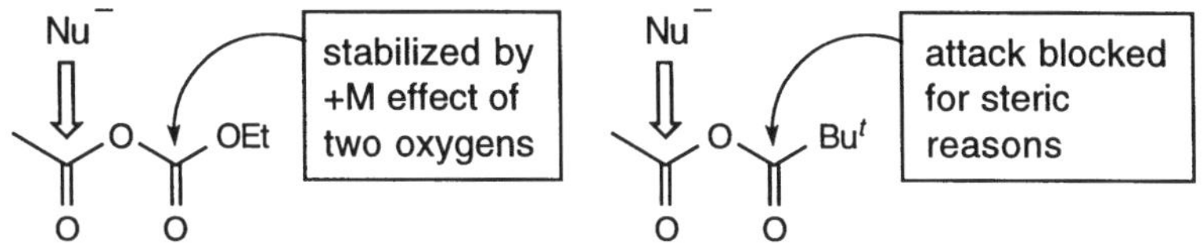

Amides via active esters   (see also Section 9.4)

Simple esters (e.g. $EtCO_2Me$) are not sufficiently reactive to be attacked by amines to form amides (RO^- is a poor leaving group). However, we can make use of –M or –I effects to help an OAr group act as a leaving group.

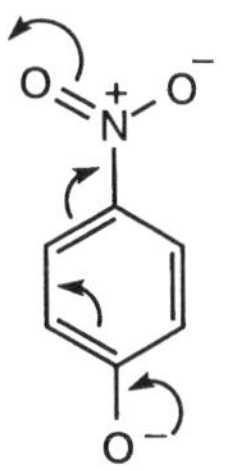

p-Nitrophenyl esters. These were once used extensively in peptide synthesis. The leaving group (ArO^-) is stabilized by the –M effect of the NO_2 group. This chemistry has been largely superseded by:

–M effect of *p*-nitro group

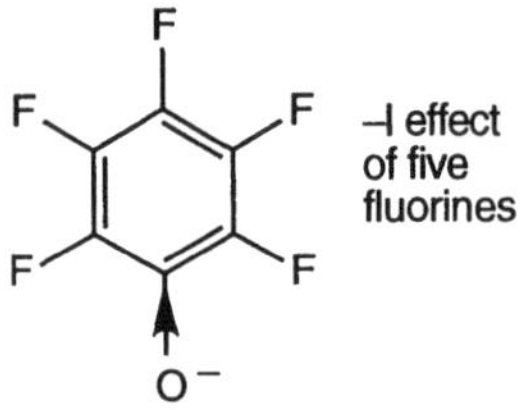

−I effect
of five
fluorines

Pentafluorophenyl esters. Here the cumulative effect of five −I fluorines helps stabilize the ArO⁻ group.

Amides via one-pot coupling reagents

Some reagents can be simply added to a mixture of an amine and an acid:

Carbodiimides. Probably the most common method of forming amides is to use carbodiimides (see Chapter 9), of which dicyclohexylcarbodiimide (DCC) is the best established reagent.

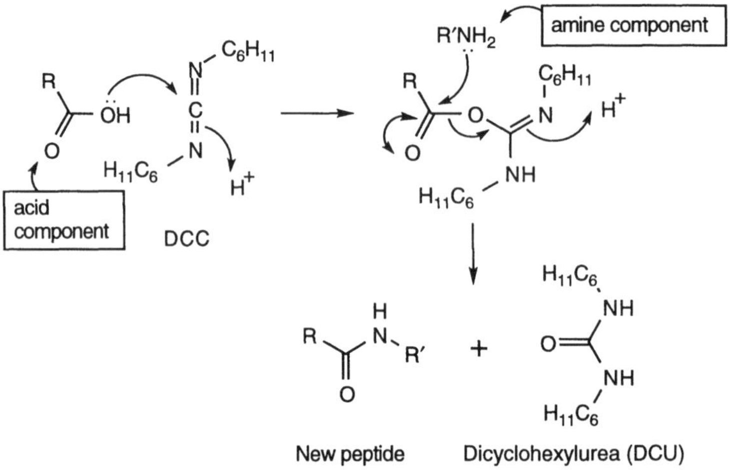

New peptide   Dicyclohexylurea (DCU)

Amazingly, the acid, amine, and DCC can be simply mixed together— hence the popularity of the method. As you can see from the mechanism, this is simply another way of generating an activated carboxylic acid derivative.

Acyl azides. These are used in fragment peptide couplings, for which racemization can be a serious risk. The low reactivity of acyl azides suppresses racemization, although peptide couplings can take several days! Classically, acyl azides are made from the simple ester (see Chapter 10):

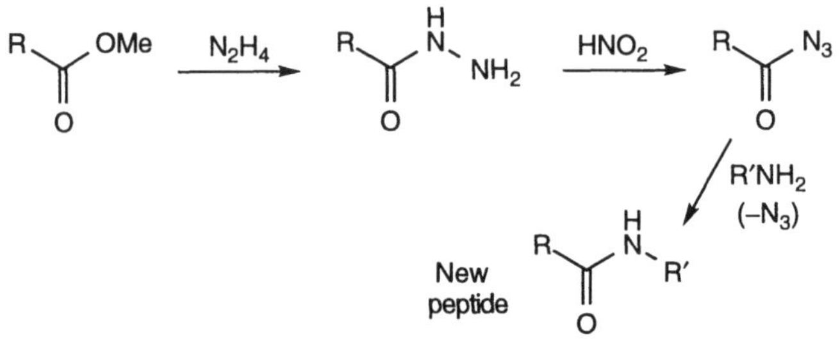

New
peptide

Phosphorus V reagents. Certain phosphorus reagents are also capable of achieving these one-pot coupling reactions; diphenylphosphoryl azide (DPPA) is one of the most widely used.

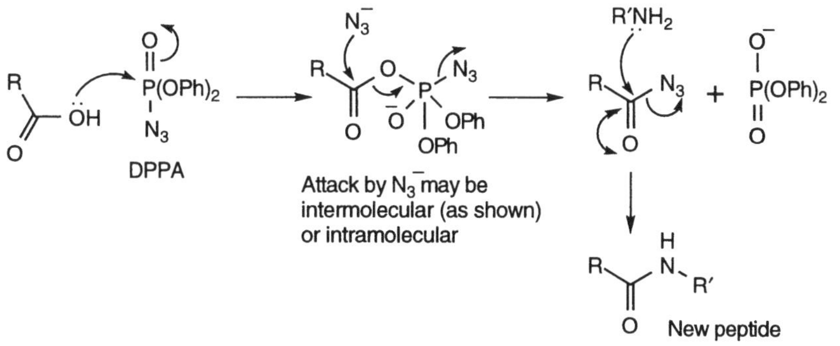

DPPA

Attack by N_3^- may be intermolecular (as shown) or intramolecular

New peptide

Racemization of activated peptides involves the following mechanism:

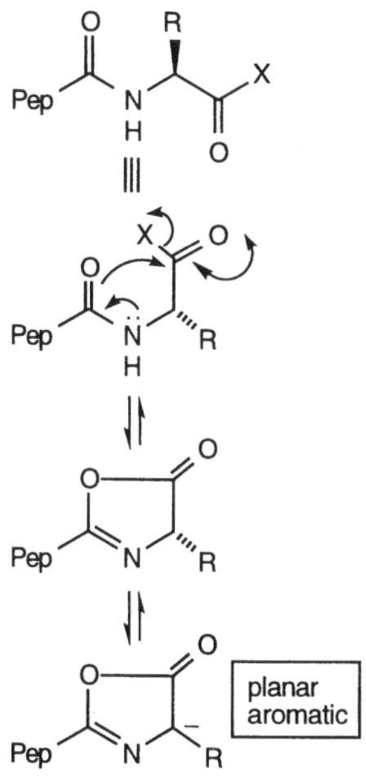

planar aromatic

This mechanism can only operate if the carboxylic acid derivative is highly activated—remember that the amide oxygen is a weak nucleophile.

The reason why a wide range of reagents has been developed for amide bond formation is that highly reactive carboxylic acid derivatives can undergo unwanted side-reactions (e.g. racemization); the more sluggish reagents are therefore often preferred.

7.3 Reactions of amides

Hydrolysis

Hydrolysis requires very forcing conditions, since the amide bond is so inert. The powerful OH⁻ nucleophile or protonation of the amide with strong acid are required before the resonance stabilization can be overcome.

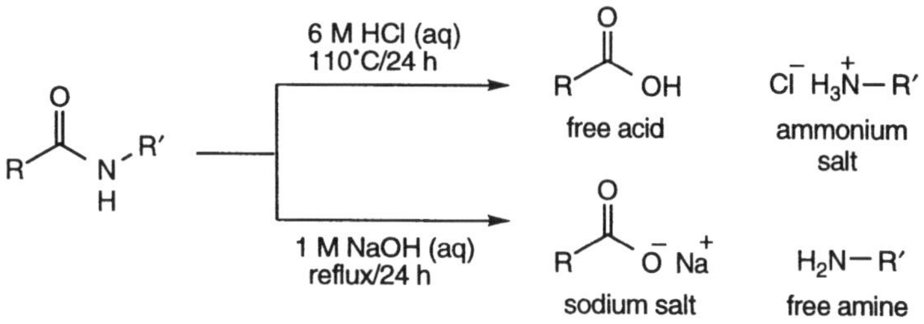

6 M HCl (aq)
110°C/24 h

free acid · ammonium salt

1 M NaOH (aq)
reflux/24 h

sodium salt · free amine

Hydrazinolysis

Other powerful nucleophiles can react with amides—hydrazine is an example.

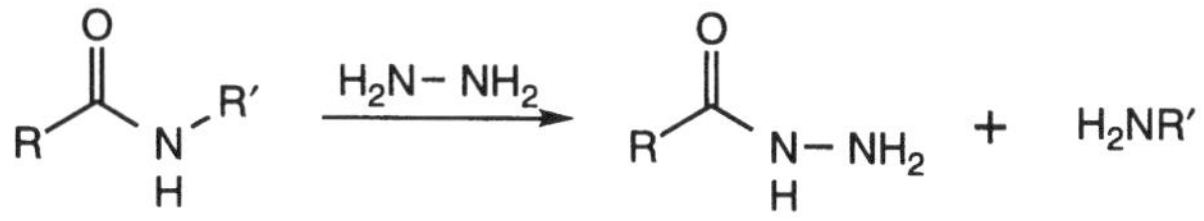

The Vilsmeyer reaction

The inert nature of the amide bond is exemplified by the widespread use of dimethylformamide (Me_2NCHO) as a solvent in organic synthesis. But with the highly reactive $POCl_3$, DMF will act as a nucleophile on oxygen, leading to the Vilsmeyer reaction discussed in Chapter 5.

The Hoffmann rearrangement

This reaction results in the loss of 'CO' from an amide and is a useful way of making some amines.

An alternative mechanism is if (X) loses Br⁻ to give a nitrene:

Attack of water on the isocyanate (Y) is acid catalysed, requiring initial protonation of the 'imine' nitrogen.

7.4 Summary

- Amides are relatively stable carbonyl derivatives, due to conjugation. This also leads to planarity of the amide bond, and (weak) nucleophilicity on oxygen:

Treatment of amides with LiAlH₄ proceeds without cleavage of the C–N bond to give amines:

Q. Can you provide a plausible mechanism ?

- Amides are usually prepared by reacting an amine with a carboxylic acid derivative.

i.e.

Acid chlorides (X = Cl) or anhydrides (X = OCOR″) are common, but acids can be activated *in situ* by diimides or some phosphorus V reagents.

- Amides are relatively unreactive, but can undergo attack on the C=O by powerful nucleophiles (e.g. LiAlH₄ or in hydrolysis). Less common reactions are the Vilsmeyer reaction and the Hoffmann rearrangement.

8 Nitriles

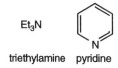

The old-fashioned name for nitriles was cyanides—but fortunately alkyl cyanides (alkanonitriles) are much less toxic than the inorganic cyanides like KCN. However, KCN and HCN are excellent sources of cyanide for introduction into organic molecules, and this is one of the reasons for the importance of nitriles in synthetic organic chemistry. But before looking at the synthesis and reactions of nitriles, we ought first to look at their reactivity in relation to other organonitrogen functional groups.

8.1 Properties of nitriles

Nitriles possess one σ- and two π-bonds between the C and N atoms. They therefore have some of the triple bond features of alkynes and some of the C=N features of imines (cf. Chapter 5).

It is worth pointing out immediately that nitriles are (perhaps surprisingly) rather unreactive. For example, acetonitrile (CH_3CN, systematically called ethanonitrile) is widely used as a solvent for organic transformations. It is readily removed after a reaction (boiling point 82°C), and has the additional feature of being water soluble (as well as being miscible with most organic solvents).

We might expect nitriles (two π-bonds) to be rather more reactive than imines (which are hydrolysed by water), so why are they so unreactive? The main reason is the low basicity of the sp-hybridized nitrogen of nitriles (compare this with the acidity of alkanes versus alkynes!)

For the same reason, triethylamine (pK_a of 9) is more basic than pyridine (pK_a of 5).

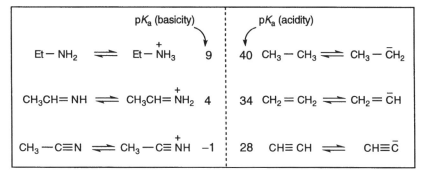

Et_3N

triethylamine   pyridine

Basicity is most conveniently expressed as the pK_a of the protonated form. Remember, pK_a is the pH at which there are equal amounts of protonated and deprotonated species (see Chapter 1).

Most reactions of imines with nucleophiles occur under acid-catalysed conditions; only very powerful nucleophiles can attack imines without such catalysis, because N dislikes carrying a negative charge.

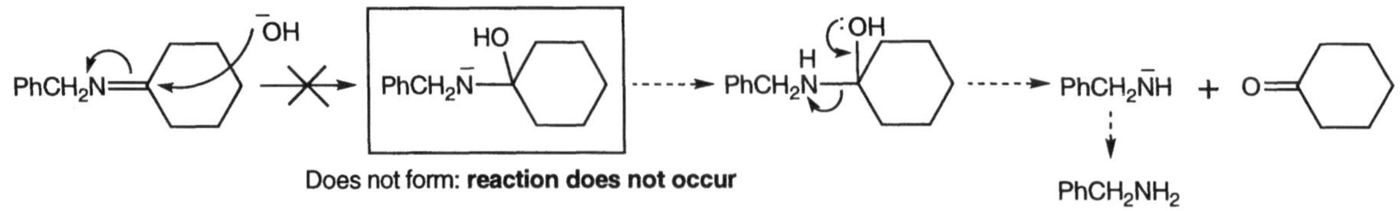

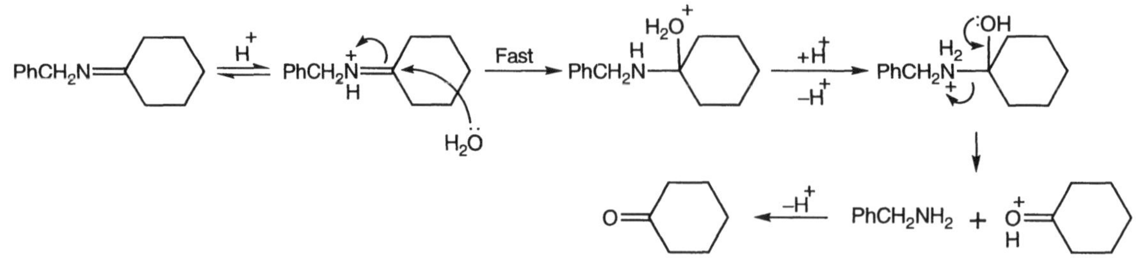

Does not form: **reaction does not occur**

Imines are stable to aqueous base, but are hydrolysed by aqueous acid. At pH 4, plenty of protonated imine is formed in the initial equilibrium, and mild nucleophiles can attack the $\delta+$ carbon of the weak π-bond. In contrast, nitriles cannot be protonated at a pH that is convenient for most organic reactions, and unless very powerful nucleophiles are employed, the nitrogen is unwilling to become negatively charged. So nitriles are relatively inert.

Nevertheless, nitriles are at the same oxidation level as carboxylic acids ($RC\equiv N$ has C with three bonds to N-heteroatoms, RCO_2H has C with three bonds to O-heteroatoms). So it is sometimes convenient to use a nitrile in place of a carboxylic acid, and then carry out hydrolysis at the end of the synthesis.

8.2 Formation of nitriles

By far the commonest way to make nitriles is to use inorganic cyanide (e.g. KCN) to introduce the $C\equiv N$ unit intact—a special version of this is the Strecker synthesis which employs HCN. It is also possible to turn primary amides into nitriles.

Tosyl (Ts):

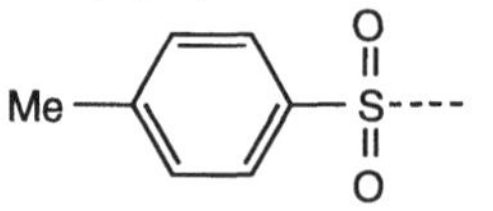

Mesyl (Ms):

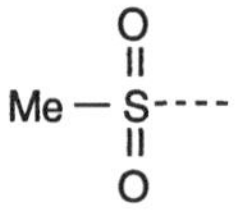

^-CN itself is perhaps surprisingly stable, but the two π-bonds to N can be strongly polarized to spread out the charge.

S_N2 reactions using inorganic CN^-

These reactions are simplicity itself, so long as you are happy handling the highly toxic (but very cheap) inorganic cyanides like KCN and NaCN. Most good leaving groups can be displaced by the cyanide, which is quite a good nucleophile (about the same as bromide).

$$R-X + KCN \longrightarrow R-CN + KX$$

$$X = Cl, Br, I, OTs, OMs$$

As a new C–C bond is formed, and the leaving groups are quite stable, these reactions proceed very readily. In contrast, ^-CN itself is a poor leaving group, requiring cleavage of a strong C–C bond. Therefore, the above reactions are very easy to do and they constitute one of the simplest ways of introducing one extra carbon into an organic molecule.

Attacking carbonyls with $^-$CN

Again, this is easy:

e.g.

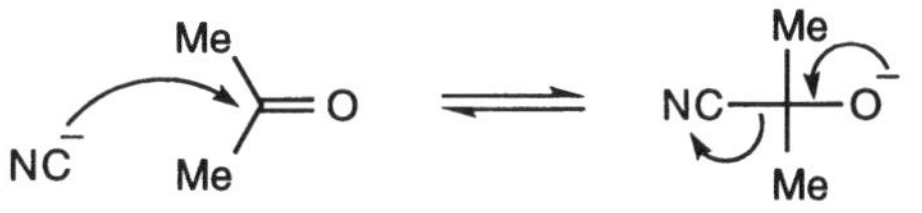

However, this equilibrium lies to the left, because the product has a negative charge on oxygen, and has had to lose a rather stable C=O π-bond. If protons are added, the equilibrium is driven to the right, but the acid reacts with KCN to give HCN! Hydrogen cyanide itself can be used directly (turning your fume cupboard into a potentially lethal 'gas chamber'!), or generated *in situ* by slow addition of acid to the KCN solution. In either case, the cyanohydrin (X) can be formed.

Actually, this equilibrium lies only just over to the right for most aldehydes. Less reactive (more stable) carbonyl compounds such as esters and many ketones usually remain unchanged; aldehydes (being more reactive) usually give good yields of cyanohydrins.

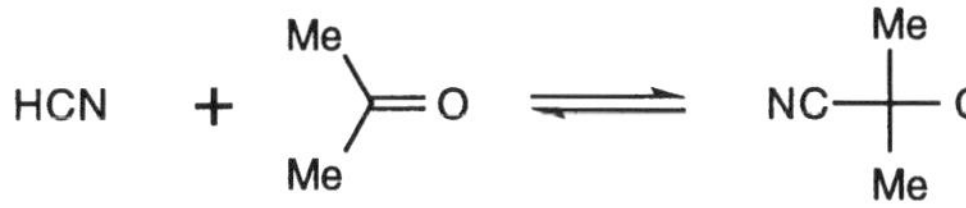

This equilibrium is exploited in a very effective synthesis of α-amino acids—the **Strecker synthesis**. If the cyanohydrin formation is carried out in the presence of ammonia with the acidity close to pH 7, then the following reaction occurs:

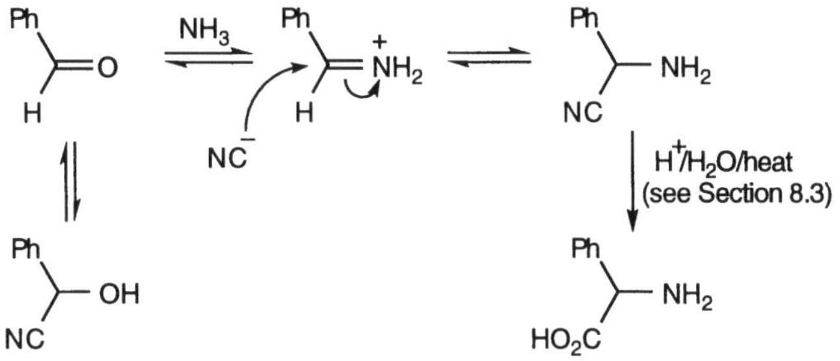

The pH is critical—too acidic, and there is no free NH_3 (only ammonium ions)—too basic, and the imine won't react (since it must be protonated before cyanide can attack). But Strecker found (by chance) that a mixture of KCN, NH_4Cl, and aqueous ammonia was naturally buffered to about the correct pH, so the reaction is very simple. The strength of the C–N bond pulls the equilibrium over to the α-aminonitrile, hydrolysis of which gives a convenient route to α-amino acids (see Section 8.3).

This route to nitriles is also valuable because of the ease with which simple **amides** can be prepared (Section 8.3):

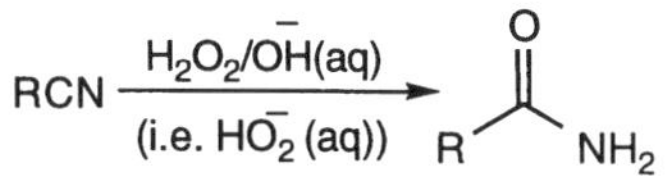

We will next look at the reverse of this 'partial' hydrolysis...

Nitriles from primary amides

If we simply look at their molecular formulae, we can see that loss of water from a primary amide could give a nitrile. This is in fact achievable quite easily...but rather powerful reagents are necessary:

e.g.

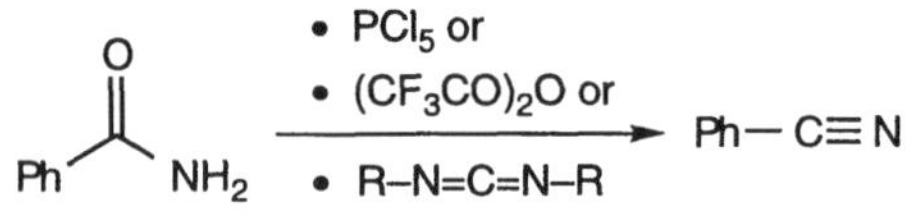

In all cases, the mechanism involves attaching a powerful electrophile to oxygen, and making it a good leaving group:

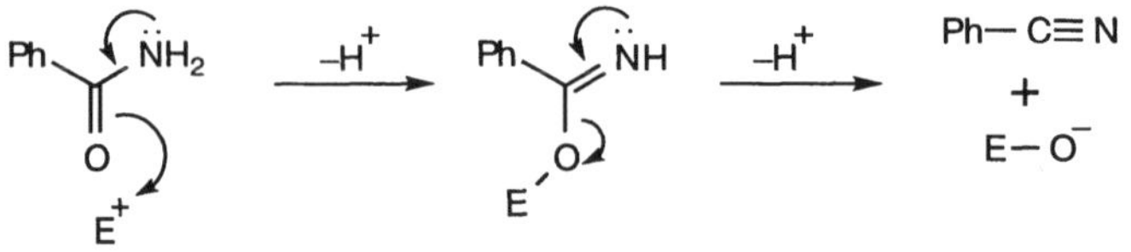

This method is very useful, provided the rest of the molecule isn't destroyed by the reagent!

8.3 Reactions of nitriles

The four main types of reaction are hydrolysis, nucleophilic attack, enolate chemistry (including the benzoin condensation), and reduction.

Hydrolysis

We have commented that nitriles are rather unreactive, and yet we have already referred to their hydrolysis. This is a very useful reaction because nitriles can be prepared rather easily. Hydrolytic conditions conditions need to be forcing—strong acid or a powerful nucleophile is required (similar to the amide hydrolysis described in Chapter 7):

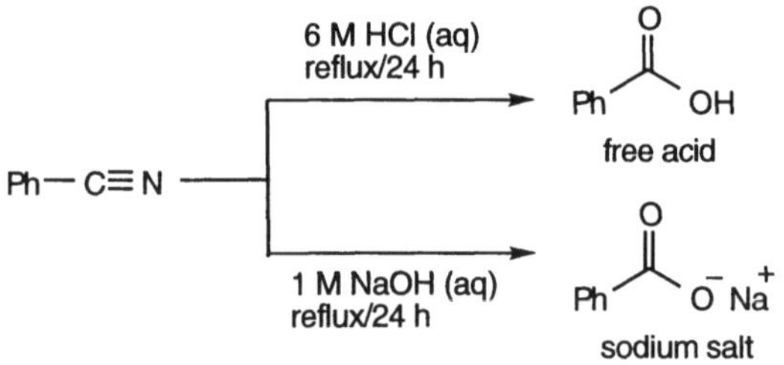

Each of these methods has a useful variant...under acid hydrolysis, the ester can be obtained in a single synthetic step:

e.g.

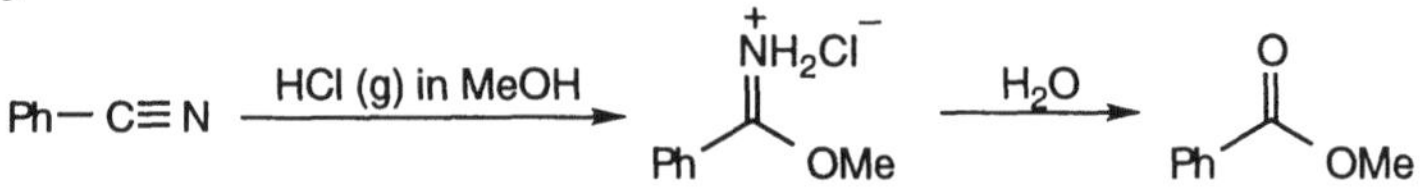

For alkaline hydrolysis, the conversion to the acid **must** necessarily proceed via the amide; it is often possible to stop the hydrolysis at this stage by using aqueous 2 M NaOH mixed with hydrogen peroxide. The more nucleophilic HO_2^- anion acts as the initial nucleophile, allowing the use of milder conditions in which the amide (being slightly less reactive than the nitrile) can be isolated.

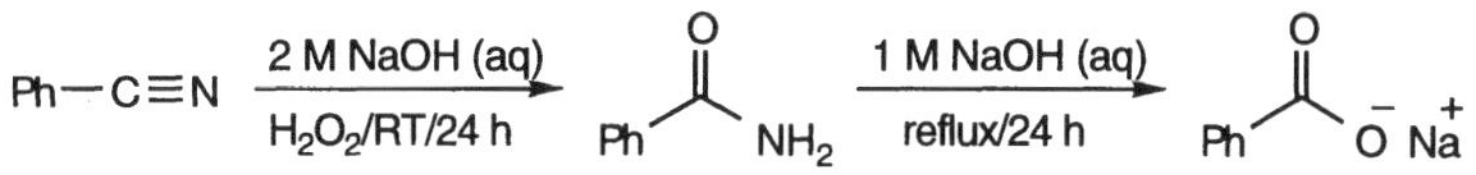

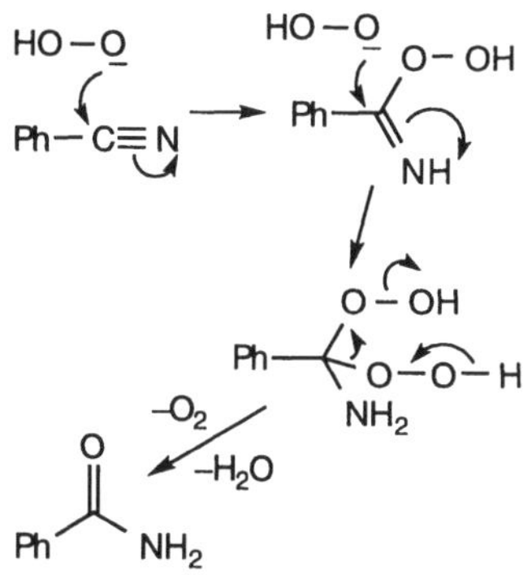

Hydrolysis (strictly speaking, solvolysis) can also be achieved under strongly acidic conditions (conc. H_2SO_4!) by trapping a carbon-cation with the weakly nucleophilic lone pair on a nitrile nitrogen. This 'Ritter' reaction requires an alcohol that can form a stabilized cation (2°, 3°, or benzylic):

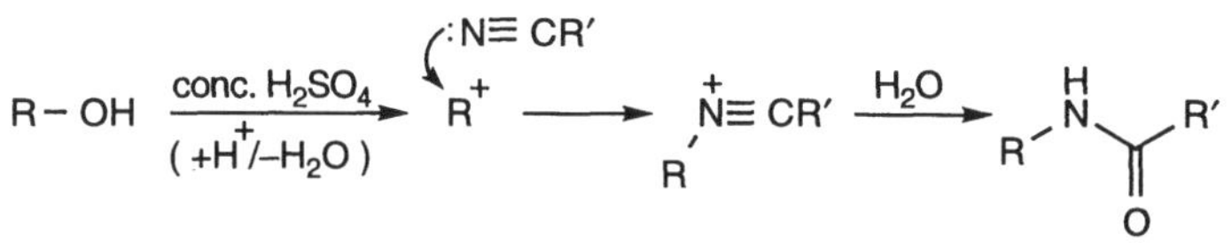

Nucleophilic attack

Because nitriles are relatively inert, powerful nucleophiles are needed to attack them (unless they can be protonated, as we saw with acidic hydrolysis). In all cases, the reaction is essentially:

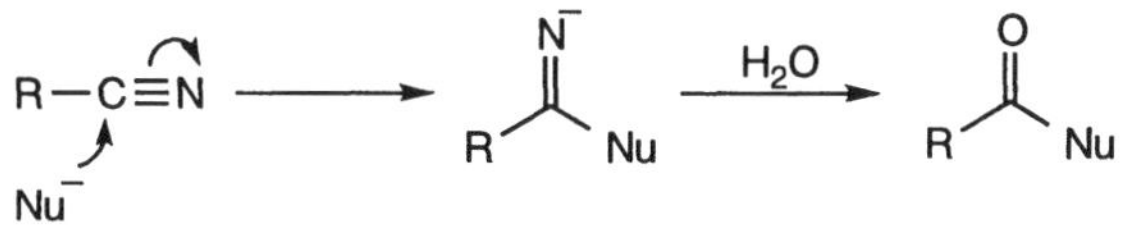

The most obvious powerful nucleophiles are Grignard reagents; the magnesium Grignards (RMgX) are the most commonly used. RLi reagents tend to deprotonate nitriles (see below), whilst organocopper reagents are not sufficiently reactive. As esters always react further with Grignard reagents (see margin), this offers a neat route to ketones, as the nitrile is attacked only once.

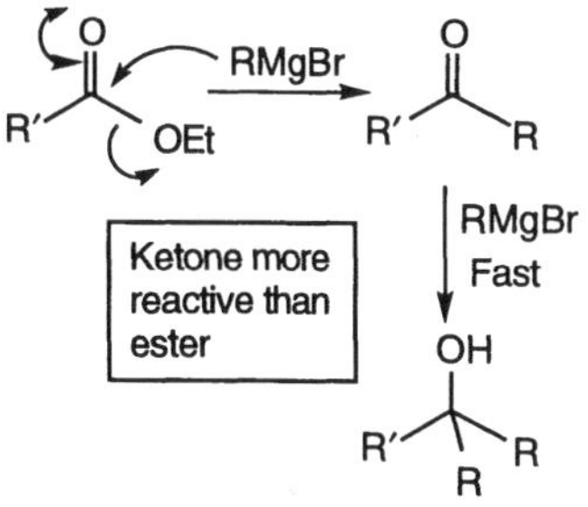

e.g.

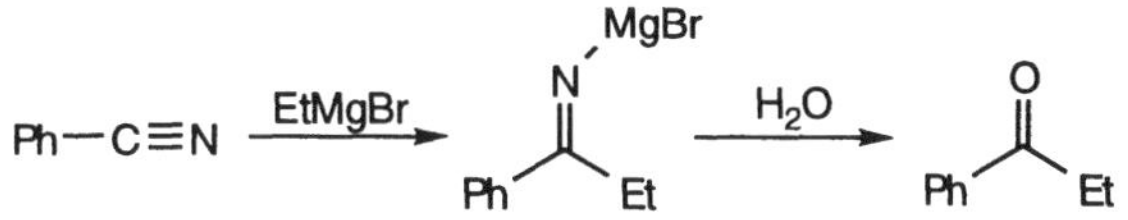

'Enolate' chemistry

With powerful bases (e.g. LDA), nitriles can be deprotonated, to give the nitrogen equivalent of an ester enolate.

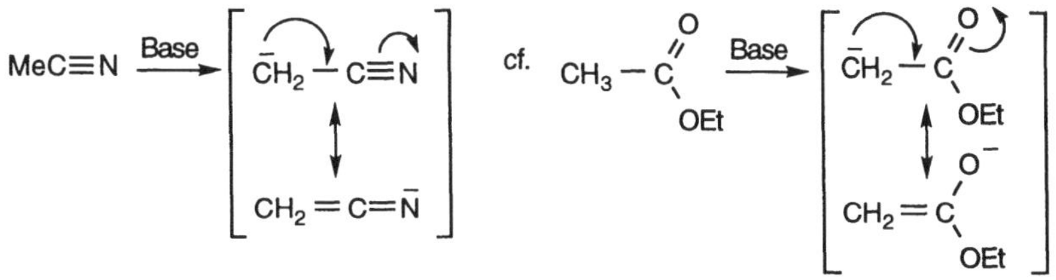

These anions can be quenched with a wide range of electrophiles, which usually react at carbon (rather than at nitrogen):

The mechanism for the partial hydrolysis of nitriles with alkaline peroxide relies upon the nucleophilicity of the HO_2^- anion:

Ketone more reactive than ester

RMgBr
Fast

Nitrile anions versus ester enolates
The electronegativity of oxygen means it is about 10^8 times better at holding negative charge than nitrogen:

$$EtOH \rightleftharpoons Et\bar{O} + H^+ \qquad pK_a \approx 16$$

$$EtNH_2 \rightleftharpoons Et\bar{N}H + H^+ \qquad pK_a \approx 24$$

The sp^2 nitrogen of a nitrile 'enolate' versus sp^3 oxygen of ester enolate offsets this by around 10^4, so:

$$MeCN \rightleftharpoons \bar{C}H_2CN + H^+ \qquad pK_a \approx 20$$

$$H_2C=C=\bar{N}$$

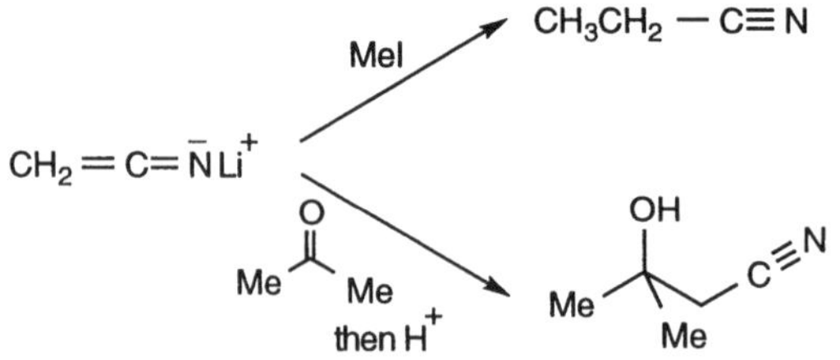

However, these reactions are rarely used synthetically because they are generally not very clean (the nitrile 'enolate' can act as a base too). But nitriles **do** feature strongly in three types of useful enolate chemistry:

(a) Nitriles offer excellent **additional** stabilization for enolates or other delocalized anions:

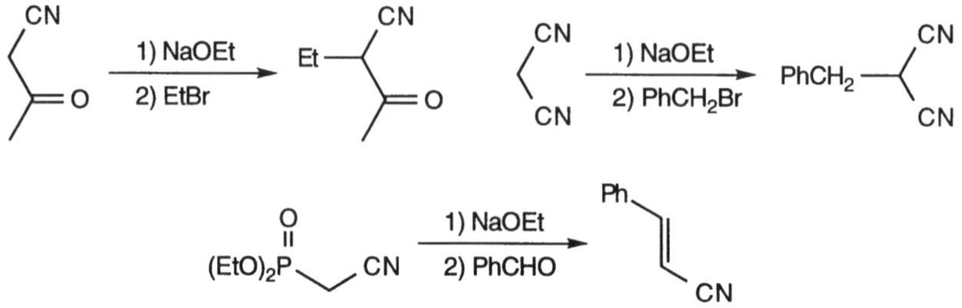

(b) Nitrile 'enolates' can be very effectively trapped (even using a mild base) in intramolecular reactions:

Cyclizations
Intramolecular reactions leading to five- or six-membered rings are common. The entropy term makes cyclization easy (i.e. there is a high probability that the reactive atoms will be close together), with minimal steric or ring strain. In the Thorpe cyclization, the rapid intramolecular process allows the very low anion concentration to be efficiently trapped out.

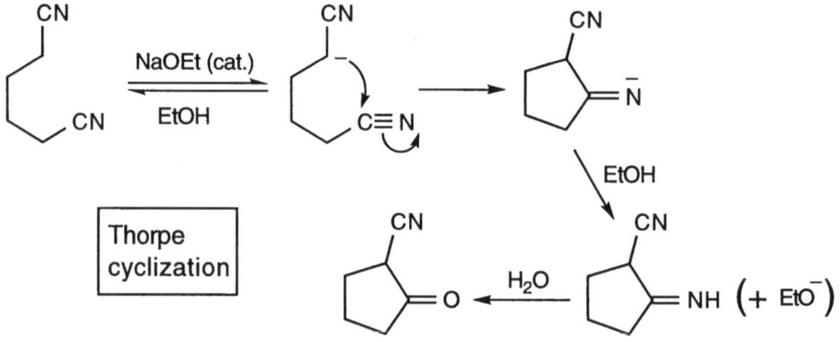

Thorpe cyclization

(c) The benzoin condensation—although nitrile is present in neither the starting material nor the product!

The cyanide has three different functions at key steps in the mechanism:
• Nucleophile
• –M group
• Leaving group

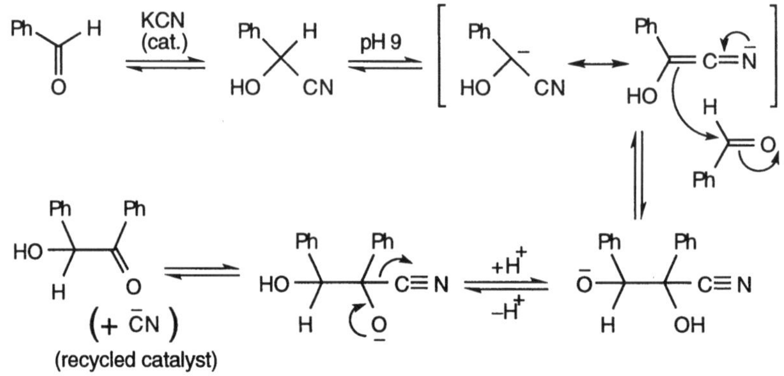

This amazing set of equilibria is all pulled over to the product (formation of new C–C bond). The success of the reaction depends critically on the pH, on the reactivity of the aldehyde (which can't undergo many other reactions), and is helped by the fact that it is a self-condensation. The wider application is limited, but it is a superb example of cyanohydrin and nitrile 'enolate' chemistry all rolled into one!

Reduction of nitriles

Mild reducing agents don't affect nitriles, but the following work well:

$$R-C\equiv N \xrightarrow[\bullet\ H_2/Raney\ Ni]{\bullet\ LiAlH_4} R-CH_2NH_2$$

$$R-C\equiv N \xrightarrow{DIBAL} R-\overset{-}{C}=\overset{+}{N}\ AlR'_2 \xrightarrow{\overset{+}{H}/H_2O} R-CHO$$
$$H$$

The last reduction is brilliant—even excess DIBAL won't over-reduce nitriles, making this an excellent route to aldehydes.

8.4 Summary

- Nitriles are relatively inert.
- They are usually prepared using inorganic cyanide as the source of $^-$CN:

$$R-X \xrightarrow{\overset{-}{C}N} R-CN$$

$$\underset{R'}{\overset{R}{>}}=O \xrightarrow{\overset{-}{C}N/NH_3} \underset{R'}{\overset{R}{>}}\underset{CN}{\overset{NH_2}{<}}$$

$$\underset{R'}{\overset{R}{>}}=O \xrightarrow{\overset{-}{C}N} \underset{R'}{\overset{R}{>}}\underset{CN}{\overset{OH}{<}}$$

$$\underset{R}{\overset{O}{\|}}\underset{NH_2}{} \xrightarrow[agent]{Strong\ dehydrating} R-C\equiv N$$

- The main reactions of nitriles are:

(a) Hydrolysis $\quad R-C\equiv N \xrightarrow[\text{or}\ \overset{-}{O}H/H_2O/heat]{\overset{+}{H}/H_2O/heat} R-CO_2H$

$R-C\equiv N \xrightarrow[2)\ H_2O]{1)\ R'OH/c.\ H_2SO_4} \underset{O}{\overset{R}{>}}\overset{H}{\underset{}{N}}R'$

(b) Nucleophilic attack $\quad R-C\equiv N \xrightarrow[2)\ H_2O]{1)\ R'MgBr} \underset{R}{\overset{R'}{>}}=O$

$R-C\equiv N \xrightarrow{HO_2^-} \underset{O}{\overset{R}{>}}NH_2$

(c) Enolate chemistry $\quad RCH_2-C\equiv N \xrightarrow{LDA} R\overset{-}{C}H-C\equiv N \leftrightarrow RCH=C=\overset{-}{N} \xrightarrow{E^+} R-\underset{E}{\overset{|}{C}H}-C\equiv N$

(d) Reduction $\quad R-CHO \xleftarrow[2)\ \overset{+}{H}/H_2O]{1)\ DIBAL} R-C\equiv N \xrightarrow[H_2/Raney\ Ni]{LiAlH_4\ or} R-CH_2NH_2$

LiAlH$_4$ fully reduces all π-heteroatom systems:

$$R-CO_2R' \longrightarrow RCH_2OH + HOR'$$

$$\underset{R}{\overset{O}{\|}}\underset{H}{\overset{}{N}}R' \longrightarrow RCH_2NHR'$$

$$R-C\equiv N \longrightarrow RCH_2NH_2$$

DIBAL/cyanohydrin methodology can be used to homologate sugars (i.e. add one CHOH unit): e.g.

Pentose $\xrightarrow[2)\ DIBAL]{1)\ HCN}$ Hexose

9 Urethanes, ureas, imides, and diimides

The functional groups in this chapter are all of the following type:

$$R-X-Y-Z-R' \quad \text{(where X, Y, Z are N, O, or C=O)}$$

They encompass the most important remaining nitrogen-containing functional groups...except for those containing N–N or N–O bonds, which are covered in Part C. We will briefly look at the synthesis of these functional groups, and at their most important chemistry.

9.1 Urethanes

Urethanes (or alkoxycarbonylamines) are of enormous importance as readily removable N-protecting groups:

$$\text{(reaction scheme)} \xrightarrow[\text{heat}]{6\ M\ HCl} EtOH + CO_2 + H_2NR$$

These hydrolytic conditions are still pretty harsh—but by changing the alkyloxy group, it is possible to engineer N-protecting groups that can be removed under a range of mild conditions.

Urethane	Abbreviation	Structure	Method of removal			
			TFA	H₂/Pd–C	[piperidine]/DMF	HBr/TFA
t-Butyloxycarbonyl	Boc	Bu^t structure	✓	X	X	✓
Benzyloxycarbonyl	Z	Ph structure	X	✓	X	✓
Fluorenyloxycarbonyl	Fmoc	structure	X	X	✓	✓

The three sets of conditions for deprotection (acidic, neutral, and basic) allow the selective removal of one type of protecting group in the presence of any of the others. This 'orthogonal' protection feature can be exploited in the synthesis of complex peptides.

Synthesis of urethanes

Simply react the amine with the appropriate acid chloride:

e.g.

Ethyl chloroformate

These 'acid chlorides' are actually chloroformates, which are themselves prepared from the alcohol and phosgene:

e.g.

Five- or six-membered cyclic urethanes (oxazolidines) can be prepared directly from the amino alcohol, or via the acyclic urethane:

e.g.

oxazolidine

Reactions of urethanes

There is only one really important reaction for urethanes—their easy hydrolysis. We've already discussed the importance of this for N-protection, but it is also worth noting that urethanes are hydrolysed much more readily than amide bonds under acidic conditions.

e.g.

The **Evans auxiliary** uses a chiral oxazolidine:

From D. A. Evans, *Aldrichimica Acta*, 1982, **15**, 23.

From valine (amino acid)

1) Base
2) E+
3) Hydrolysis

E introduced with high stereocontrol

(recovered)

Two factors drive this reaction: the N can be protonated at an accessible pH (cf. amide) and stable CO_2 is generated.

The stabilization of the urethane carbonyl by **two** +M groups makes it resistant to nucleophilic attack. Many nucleophiles will attack other functional groups in the presence of urethanes, further explaining their value as protecting groups.

e.g.

and

Very strong nucleophiles are required to attack the carbonyl group of urethanes. For example, $LiAlH_4$ is the only common reducing agent that will reduce them.

e.g.

The above 'two-way' delocalization strongly stabilizes the C=O bond, but causes the (non-carbonyl) O and N atoms to be more reactive than in esters or amides.
The unmasking of an amine protected as a urethane can follow several mechanisms depending on R...but the loss of stable CO_2 is the driving force for the generation of the free amine.

As we would expect, the C–O bond breaks in preference to the C–N bond, and the amide intermediate is more reactive (to nucleophiles) than the starting urethane, so reduction right down to the *N*-methyl derivative occurs.

9.2 Ureas

Synthesis of ureas

If two equivalents of an amine are reacted with one equivalent of phosgene, then a urea should form:

e.g.

The second step will be quite slow, as the amido acid chloride (X) is much less electrophilic than phosgene. Actually, ureas of this type are of little use, but cyclic ureas are much easier to prepare, and can be used to aid stereochemical analysis.

e.g.

| Stereochemical assignment difficult | Easy to form...and coupling correlation between H^A and H^B (J_{AB}) or NOEs between these protons can be used to show that they are *cis* |

Phosgene is pretty reactive (and toxic!). Some modern alternatives that are milder and safer are all of the type X–CO–X:

'Triphosgene' Diimidazole urea Dimethyl carbonate

Their reactivity is in the order shown; only cyclic ureas are accessible from dimethyl carbonate, whereas the highly reactive 'triphosgene' actually forms three moles of phosgene in its reaction with nucleophiles.

Reactions of ureas

Ureas are notable for their lack of reactivity (the carbonyl is stabilized by two +M nitrogens). Strong alkaline hydrolysis yields amines plus CO_2:

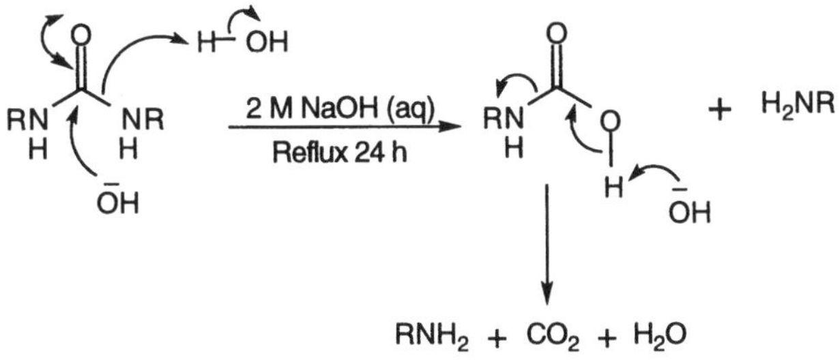
2 M NaOH (aq)
Reflux 24 h
$RNH_2 + CO_2 + H_2O$

The only other widely used reaction is dehydration. It requires very powerful reagents, but generates valuable diimides (see Section 9.4).

e.g.

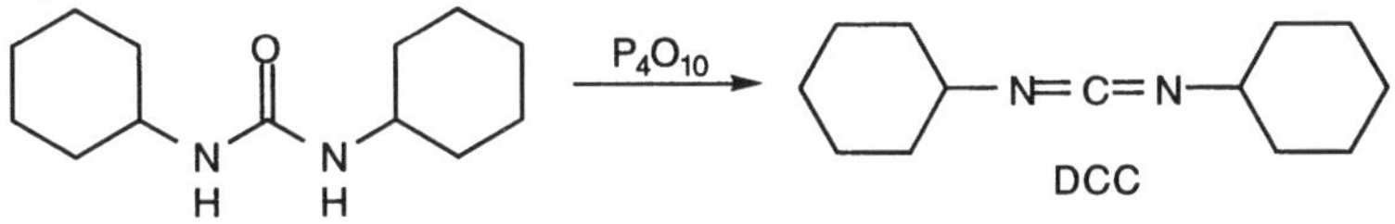
P_4O_{10}
DCC

One important group of cyclic urea derivatives is the **barbiturates**.

e.g.

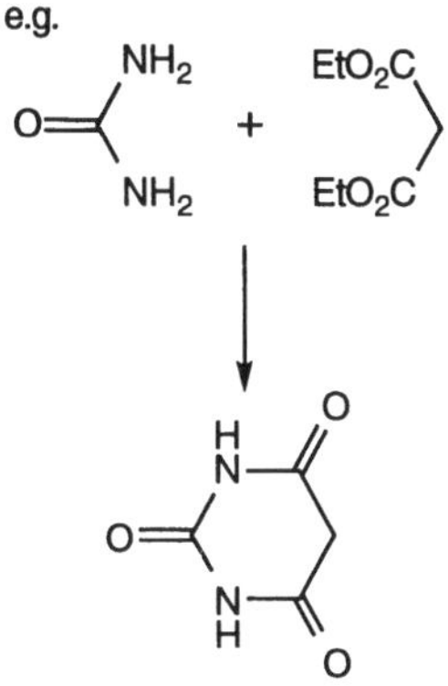

9.3 Imides

Acyclic imides occur very rarely, but five- and six-membered cyclic imides are both easy to prepare, and are of substantial synthetic value.

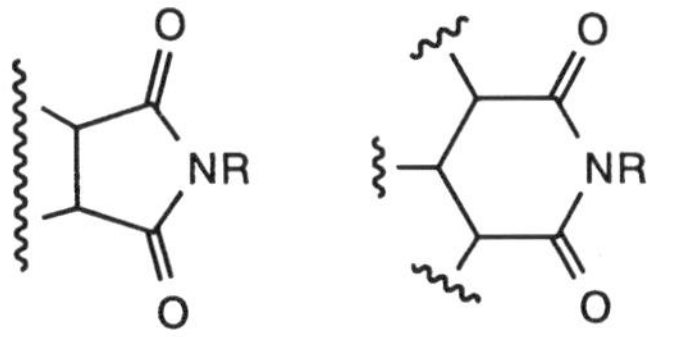

Synthesis of imides

Simply heating cyclic anhydrides with ammonia or a primary amine generates the imide.

e.g.

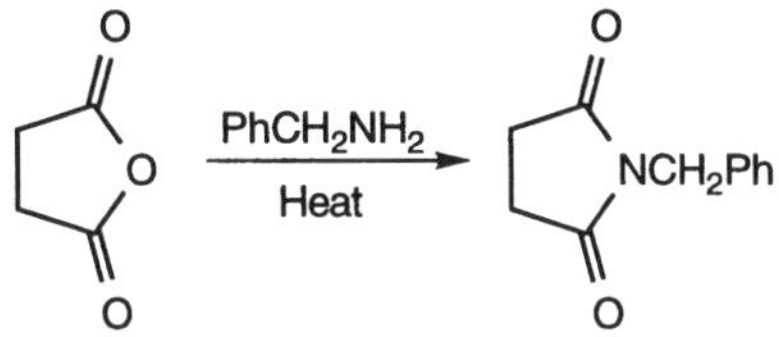
$PhCH_2NH_2$
Heat

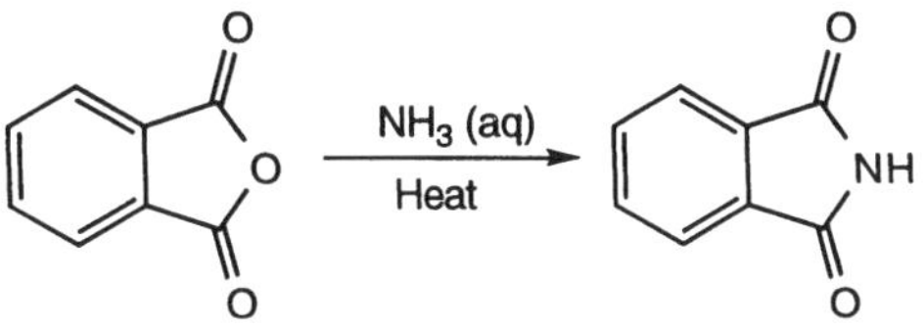
NH_3 (aq)
Heat

Cyclic anhydrides are simply formed from the diacid using almost any reagent that can form an activated carboxylic acid derivative.

e.g.

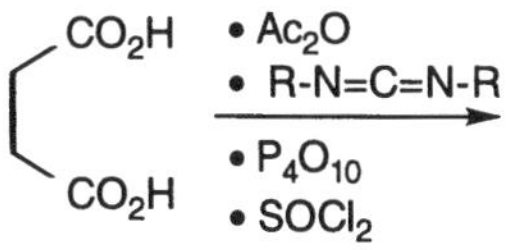
• Ac_2O
• R-N=C=N-R
• P_4O_{10}
• $SOCl_2$

The mechanism is:

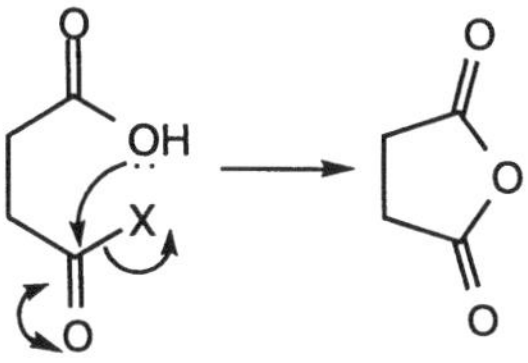

Reactions of imides

There are two main types of reaction; the first is when the imide has an NH group, as discussed in Chapter 4. When treated with a strong base (e.g. potassium hydride) the proton is removed to give a stabilized anion. The nucleophilic nitrogen can then be monoalkylated quite readily. This reaction is important because the alkylation can occur only **once**, and subsequent hydrolysis of the imide generates a primary amine. This is in contrast to the problems encountered when one tries to form primary amines by mono-alkylation of ammonia (very difficult to do efficiently—see Chapter 1). Because of this, Gabriel's reagent has been widely used as a source of 'mono-nucleophilic' ammonia, although a number of alternatives (e.g. azide) are more popular now (see Chapter 4). This is mainly because the conditions for removal (i.e. hydrolysis) of the imide unit are so harsh—strong aqueous base will do it, but hydrazine is the more usual reagent:

negative charge delocalized onto two oxygens

The formation of the stable aromatic six-membered heterocycle (X) is the driving force for the hydrazinolysis of the phthalimide ring.

The imide carbonyls are more reactive than amide carbonyls because they only have a 50% share in the nitrogen lone pair. After the reaction of one C=O with a nucleophile, the other becomes a normal amide and is less reactive.

Imides that possess an N–R group react quite efficiently **once** only with a range of nucleophiles. This feature can be exploited if the imide ring is to become part of the target molecule (rather than simply facilitating reactions on nitrogen before being removed, like Gabriel's reagent):

Mannich reaction (see Chapter 5)

See P. D. Bailey, K. M. Morgan, D. I. Smith, and J. M. Vernon, *Tetrahedron Lett.*, 1994, **35**, 7115.

Reduction with $NaBH_4$ similarly leads to monoreduction of one of the carbonyl groups but, as you would expect, $LiAlH_4$ effects reduction down to the cyclic amine.

9.4 Diimides

Synthesis of diimides

Diimides are readily made from ureas (see Section 9.2):

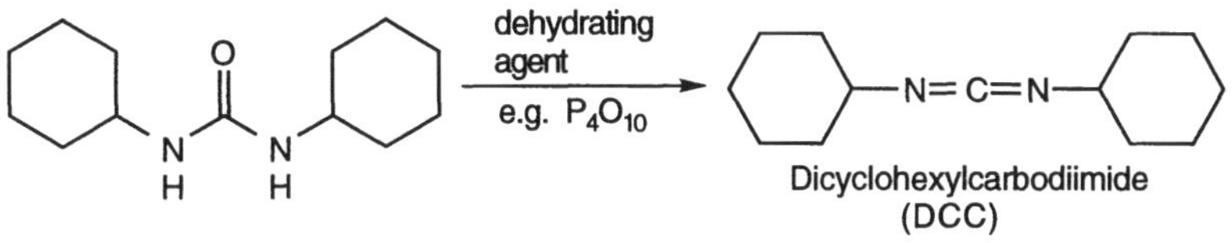

Reactions of diimides

Diimides are superb reagents for achieving one particular type of reaction; turning carboxylic acids into esters or amides. The commonest diimide that is used is the dicyclohexyl derivative (DCC, see above).

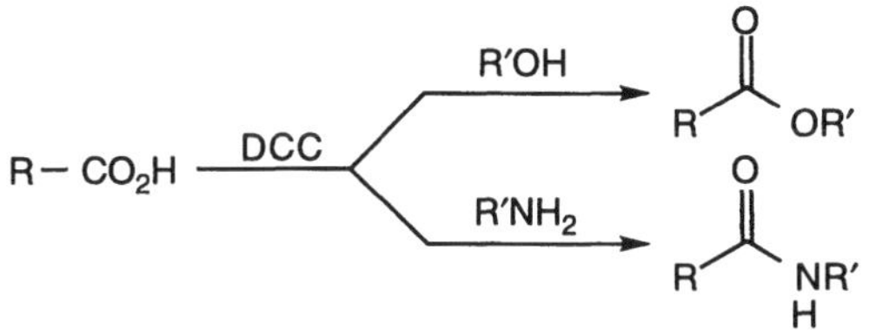

DCC is one of the most widely used reagents for preparing **peptides**. The success of DCC in peptide chemistry centres around the fact that all the reactants can be mixed in one pot, as discussed in detail in Chapter 7. The excellent activating power of DCC can lead to serious side-reactions in peptide synthesis, so moderating agents are sometimes added that generate active esters *in situ*; typical additives include *N*-hydroxysuccinimide (NHS) or *N*-hydroxybenzotriazole (HOBt):

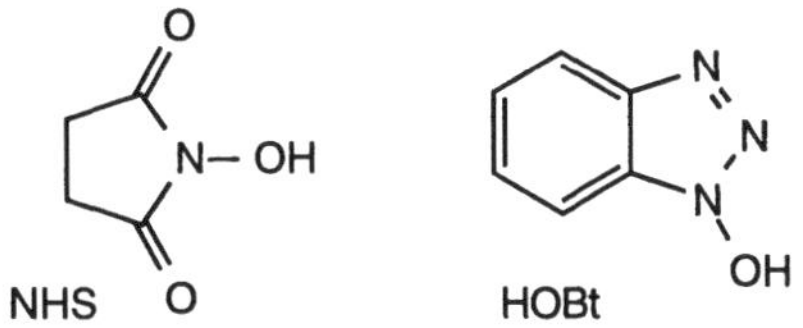

Simple esters are rarely made using DCC—there are usually easier methods available. But active esters in general are hard to make, so reaction of the acid plus alcohol in the presence of DCC is a very useful method. The pNP-esters and PFP-esters are good examples, which are themselves used in peptide synthesis (again, see Chapter 7).

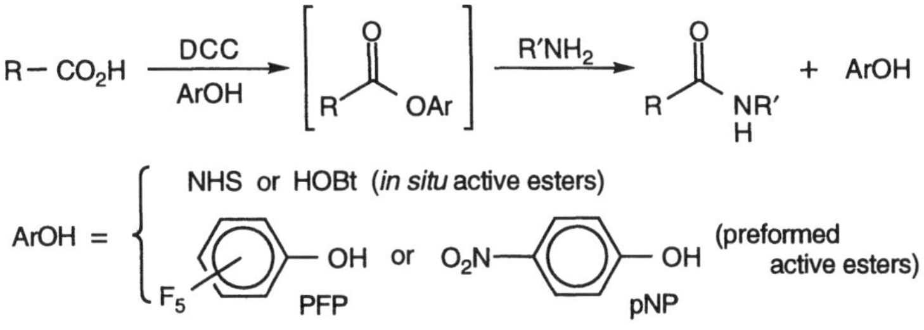

If a carboxylic acid is reacted with half an equivalent of DCC, in the absence of other nucleophiles, then the acid can itself act as a nucleophile, and this is a useful way of making acid anhydrides...which are also used in peptide synthesis (see Chapter 7).

So successful are diimides in peptide synthesis that derivatives other than DCC have been developed. Perhaps one of the most useful is the water-soluble diimide EDC (ethyl-(diethyl-aminoethyl)carbodiimide), in which the extra amino group makes all EDC-related compounds soluble in aqueous acid, and hence removable by a simple acid wash.

9.5 Other unsaturated nitrogen-containing groups

If you play around with pen and paper, many more nitrogen-containing functional groups can be drawn. The most important ones are discussed in this book, whilst the properties of many of the others can be largely inferred from chemical common sense, and by comparison with other N-containing functional groups. Some other functional groups include:

amidines imidines guanidines imidate esters

9.6 Summary

- Urethanes are invaluable for protecting amines whilst other functional groups are being modified.

The conditions required for deprotection can be controlled by the choice of R.

- Imides are very useful for preparing **primary amines**.

- Diimides are one of the best types of reagent for the coupling of carboxylic acids to amines. This reaction is widely used in peptide synthesis (often in conjunction with urethane protection).

$$RCO_2H + R'NH_2 \xrightarrow[(-H_2O)]{DCC}$$

Nitrogen compounds with N–O or N–N bonds

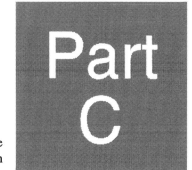

In Parts A and B we focused on organonitrogen compounds in which the nitrogen was bonded only to carbon or hydrogen. Although this yields a rich chemistry, the number of nitrogen-containing functional groups increases enormously if we also consider N–N and N–O bonds.

$R-NH-NH_2$	Hydrazine	$R-\overset{+}{N}{=}\overset{-}{N}{=}\overset{-}{N}$	Azide
$R\diagdown\overset{H}{\underset{\|}{N}}-NH_2$	Hydrazide	$R-CH{=}N-OH$	Oxime
		$R-N{=}O$	Nitroso
		$R_2N-N{=}O$	N-Nitroso
$R-CH{=}N-NH_2$	Hydrazone	$R-NO_2$	Nitro
$R-N{=}N-R$	Diazo	$R_3\overset{+}{N}-\overset{-}{O}$	N-Oxide

The reactivity of these compounds varies enormously, and we will consider some of the factors that influence this before looking at specific functional groups in more detail.

Strength of neutral N–N and N–O bonds

The main bonds we need to consider are N–N, N–O, N=N, and N=O. In general, carbon forms strong bonds with itself. In contrast, bonding is weaker between a pair of atoms which possess lone pairs because of lone pair–lone pair repulsion...so N–N and N–O systems are generally less stable than functional groups lacking this interaction.

The presence of π-bonds between the heteroatoms often provides a mechanism by which such compounds can react. For example, the N–O bond may be thermodynamically quite weak, but (perhaps surprisingly) such compounds are relatively stable because there is no simple mechanism by which the N–O bond can be broken. In contrast, nitroso compounds (R–N=O) are very reactive since the polarized N=O π-bond facilitates mechanisms for a wide range of reactions.

The stability of N–N and N–O systems is also strongly influenced by conjugated π-bonds. Thus oximes are stabilized by the π-bond being able to overlap with the lone pair of electrons on the oxygen:

$$\left[R-CH{=}N{\overset{\curvearrowright}{\cdot\cdot}}\ddot{O}H \longleftrightarrow R-\overset{-}{C}H-N{\overset{\nearrow}{=}}\overset{+}{O}H \right]$$

Typical bond strengths of C–C, N–N, and N–O bonds are:

C–C	345 kJ mol^{-1}
N–N	220 kJ mol^{-1}
N–O	220 kJ mol^{-1}

Charged functional groups

Some of the most important N–O-containing functional groups are zwitterionic—they possess both a positive **and** a negatively charged atom within the overall neutral group. One would normally expect charged compounds to be highly reactive, so why is that not the case with many charged N–N and N–O systems? The full reasons will become more apparent in later chapters, but some key factors can be gleaned by considering nitro compounds ($R–NO_2$):

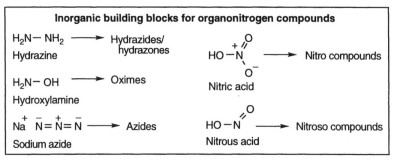

- The NO_2 functional group is strongly resonance stabilized—this stabilization is lost during any chemical transformation.
- The N and O atoms possess a full octet of electrons and have a charge which they are able to accommodate quite easily (remember that ammonium cations and alkoxide anions are relatively stable).
- There is no simple mechanism by which the zwitterion can react—this becomes apparent when you try the arrow pushing required to react with an electrophile or a nucleophile.

Inorganic counterparts

For many N–N and N–O functional groups, there is a simple inorganic analogue that contains the structural unit. These inorganic compounds are usually the source of the functional group when these organonitrogen compounds are synthesized:

Inorganic building blocks for organonitrogen compounds

$H_2N–NH_2$ ⟶ Hydrazides/hydrazones
Hydrazine

$H_2N–OH$ ⟶ Oximes
Hydroxylamine

$Na^+ \; \bar{N}=\overset{+}{N}=\bar{N}$ ⟶ Azides
Sodium azide

$HO–\overset{+}{N}(=O)(–O^-)$ ⟶ Nitro compounds
Nitric acid

$HO–N(=O)$ ⟶ Nitroso compounds
Nitrous acid

With these inorganic building blocks available, a wide range of N–N and N–O organonitrogen compounds are accessible. Of special importance is that nitrous and nitric acids can be sources of $[NO]^+$ and $[NO_2]^+$ respectively.

Nucleophiles readily become attached to the N of $[NO]^+$ to give nitroso compounds. The HNO_2 is usually generated *in situ* from $NaNO_2$ and dilute acid (e.g. HCl).

This equilibrium lies much further over to the left than that of $HNO_2/[NO]^+$. Concentrated sulphuric acid is often added as a strong acid **and** a dehydrating agent. Nucleophiles readily attack the $[NO_2]^+$ to give nitro compounds.

10 Compounds with N–N bonds

$$R-NH-NH_2$$
$$R-CH=N-NH_2$$
$$R-N=N-R$$
$$R-N=\overset{+}{N}=\overset{-}{N}$$

All of these functional groups are closely related. However, we will consider their synthesis and properties independently, although the overlap in their chemistry should become obvious.

10.1 Hydrazines and hydrazides

Synthesis

Alkylhydrazines are usually made from a suitable alkyl halide and hydrazine.
e.g.

As with the synthesis of amines from ammonia, care must be taken to prevent polyalkylation of the hydrazine.

Aryl halides can sometimes also be made in this way; the best example is the very useful 2,4-dinitrophenylhydrazine:

dinitrophenylhydrazine (DNP)

Nucleophilic aromatic substitution is only possible here because of the strongly electron withdrawing nitro groups. In their absence, displacement of the halide would not occur.

Acyl hydrazines are called hydrazides (or acid hydrazides), and they can be prepared from acid chlorides. However, esters also react smoothly with hydrazine, and this is the preferred method of preparation.
e.g.

Hydrazine (H_2N–NH_2)
Hydrazine is usually sold as its hydrate (H_2N–NH_2·H_2O) or as its hydrochloride salt. At high temperatures, it decomposes to nitrogen and ammonia:

$$3\ H_2N-NH_2 \xrightarrow{\text{heat}} 4\ NH_3 + N_2$$

Once ignited, the reaction is exothermic, and leads to a big increase in volume (x 5/3). As no co-reactant is needed, hydrazine is an ideal rocket fuel which is used extensively.

Q. Draw the mechanism of this reaction, and hence explain why chlorobenzene does not react with hydrazine.
Hint: look in Chapter 14!

Reactions of hydrazines and hydrazides

The most important property of alkyl and aryl hydrazines is their ability to form hydrazones with aldehydes and ketones. 2,4-Dinitrophenylhydrazine (otherwise known as Brady's reagent or DNP) is widely used and gives a bright orange precipitate because of the extensive conjugation in the product. The formation of this precipitate is a classical test for aldehydes and ketones, and the characteristic melting points of different DNP derivatives have been used to identify specific carbonyl compounds.

e.g.

This reaction is effectively the formation of an imine (see Chapter 5).

orange DNP derivative

This reaction is driven by loss of water, so only primary hydrazines can react, and they do so at the free $-NH_2$ end. In contrast, simple alkylation reactions lack selectivity, and tend to give mixtures of substituted hydrazines. However, the full range of mono-, di-, and trimethyl hydrazines are commercially available, so derivatives of these are often quite easy to make (e.g. MeNH–NHMe is symmetrical, so alkylation is quite efficient; in Me_2N–NH_2 only the free NH_2 is reactive).

Acyl hydrazines (hydrazides) have very strongly differentiated nitrogens. This can be exploited by reaction with nitrous acid, which forms the acyl azide:

Hydrazine-like; reactive

Amide-like; unreactive

$$2\,HNO_2 \rightleftharpoons [NO]^+ + H_2O + NO_2^-$$

To generate $[NO]^+$ it is often more convenient to treat an alkyl nitrite with acid.

e.g.

$$Bu^tO\text{–}NO + CF_3CO_2H$$

$$\downarrow$$

$$[NO]^+ + Bu^tOH$$
$$+ CF_3CO_2^-$$

The acyl azides can now be used as acylating agents, and this route to them is especially important in peptide synthesis (see Chapter 7).

10.2 Hydrazones

Synthesis

As discussed in the previous section, hydrazones are prepared from the reaction of aldehydes and ketones with hydrazines. Loss of water is an important driving force for this reaction, which follows an analogous pathway to imine formation (see Chapter 5).

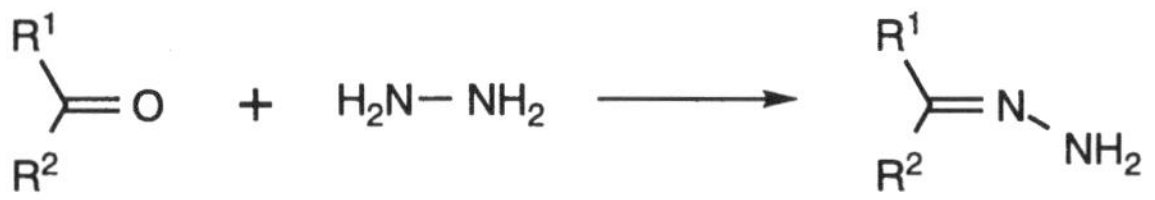

$$R^1R^2C{=}O + H_2N{-}NH_2 \longrightarrow R^1R^2C{=}N{-}NH_2$$

Reactions of hydrazones

Under conditions of vigorous acidic hydrolysis hydrazones can be forced to regenerate the carbonyl compound. However, due to conjugation of the terminal nitrogen lone pair with the π-bond, hydrazones are much less reactive than imines. One important reaction is the Wolff–Kishner reduction. Under very forcing conditions, aldehydes and ketones can be reduced to their methylene analogues:

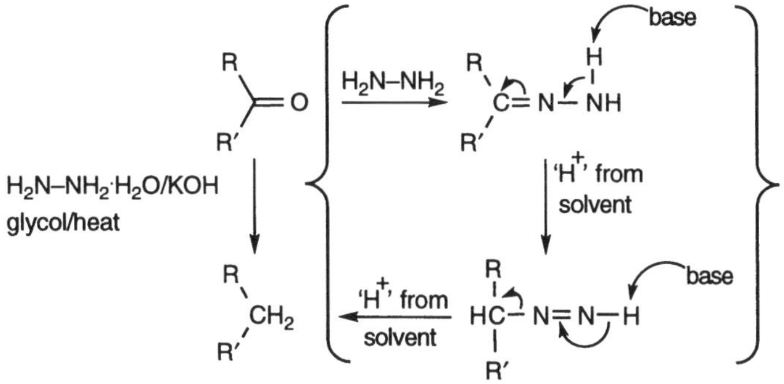

Enders' reagents
Enders has made elegant use of hydrazones in the asymmetric synthesis of chiral α-substituted aldehydes or ketones, using hydrazine reagents SAMP and RAMP, derived from (*S*)- and (*R*)-proline: e.g.

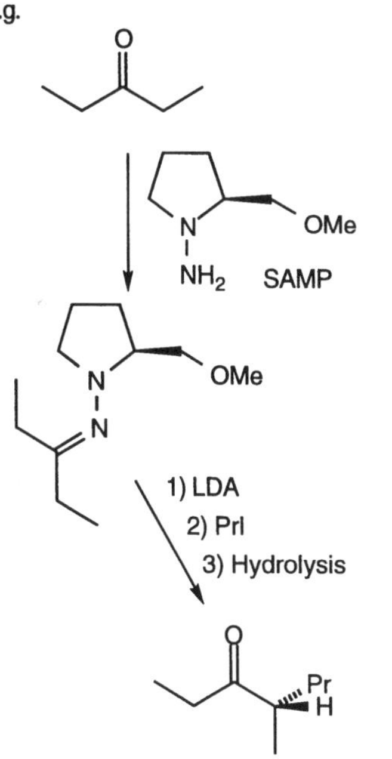

10.3 Diazo compounds

Synthesis

It is not obvious how to prepare such compounds, since a suitable 'N=N' inorganic building block is not available. They can usually be prepared only when the groups either side of nitrogen are fully substituted—otherwise the hydrazone is the preferred isomer. Thus, oxidation of certain hydrazines leads to the formation of diazo compounds.

e.g.

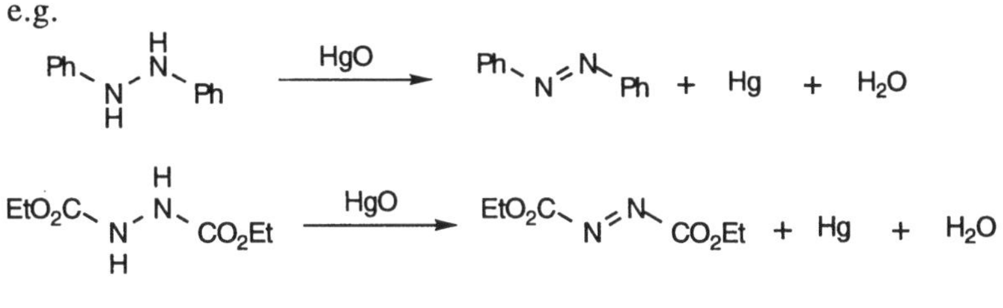

$$Ph{-}N(H){-}N(H){-}Ph \xrightarrow{\text{HgO}} Ph{-}N{=}N{-}Ph + Hg + H_2O$$

$$EtO_2C{-}N(H){-}N(H){-}CO_2Et \xrightarrow{\text{HgO}} EtO_2C{-}N{=}N{-}CO_2Et + Hg + H_2O$$

Ant alarm pheromone

Q. Suggest a mechanism.

From D. Enders *et al.*, *Organic Syntheses*, 1987, **65**, 173 and 183.

Aromatic diazonium compounds are extremely useful in aromatic synthesis:

$$ArNH_2 \xrightarrow[\text{(HNO}_2\text{)}]{[NO]^+} Ar\overset{+}{N}\equiv N$$

$$\downarrow CuX \quad \begin{array}{l} (X = Br, \\ Cl, CN) \end{array}$$

$$Ar{-}X$$
(Sandmeyer reaction)

Aliphatic diazonium compounds are similarly formed from primary amines, but usually undergo immediate loss of nitrogen because the '$-N_2^+$' is not stabilized; the corresponding alcohol is usually recovered:

$$PhCH_2NH_2 \xrightarrow[\text{(HNO}_2\text{)}]{[NO]^+}$$

$$PhCH_2{-}\overset{+}{N}\equiv N \xrightarrow{-N_2}$$

$$\left[Ph\overset{+}{C}H_2 \right] \xrightarrow{H_2\ddot{O}} PhCH_2OH$$

The classical way to prepare aromatic diazo dyes was to trap out the diazonium compounds with a carbon nucleophile:

e.g.

Reactions of diazo compounds

Diazo compounds undergo few useful reactions, and are mainly of importance because of their strong colours—used extensively in the old dye industry, and for identification of some functional groups.

10.4 Azides

Synthesis

Alkyl azides are simply prepared by nucleophilic attack of (sodium) azide on a suitable alkyl compound—usually a halide or sulphonate.

e.g.

$$PhCH_2Br \xrightarrow{NaN_3} PhCH_2{-}N{=}\overset{+}{N}{=}\overset{-}{N}$$

Importantly, the azide is only **mononucleophilic**—the azide product does not react with excess alkyl halides, and this is why primary amines are often made via the azide (see Chapter 4).

Acyl azides can be prepared from the acid chloride, or via the hydrazide.

e.g.

Reactions of azides

As indicated above, alkyl azides are commonly used as intermediates in the synthesis of primary amines. Simple catalytic hydrogenation effects the reduction very cleanly, with loss of N_2 being a powerful driving force.

e.g.

$$PhCH_2{-}N{=}\overset{+}{N}{=}\overset{-}{N} \xrightarrow{H_2/Pd{-}C} PhCH_2{-}NH_2 + N_2$$

As mentioned in Chapter 7, acyl azides are important intermediates in peptide synthesis. They also undergo a useful rearrangement reaction—the Schmidt rearrangement, which forms isocyanates:

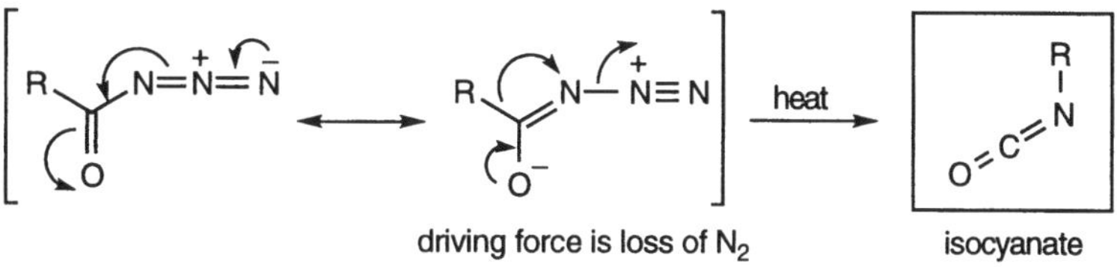

driving force is loss of N_2 isocyanate

Nitrenes
Irradiation or heating of azides gives highly reactive nitrenes:

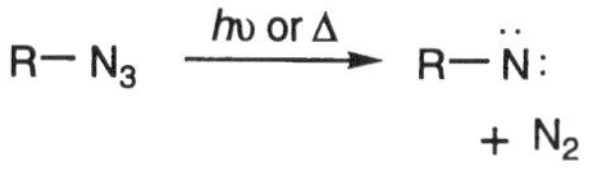

For details, see the OUP Primer *Reactive Intermediates* by C. J. Moody and G. H. Whitham.

Hydrolysis of the isocyanate with an alcohol gives urethanes (protected amines, see Chapter 9), or hydrolysis with water followed by decarboxylation gives the free amine directly:

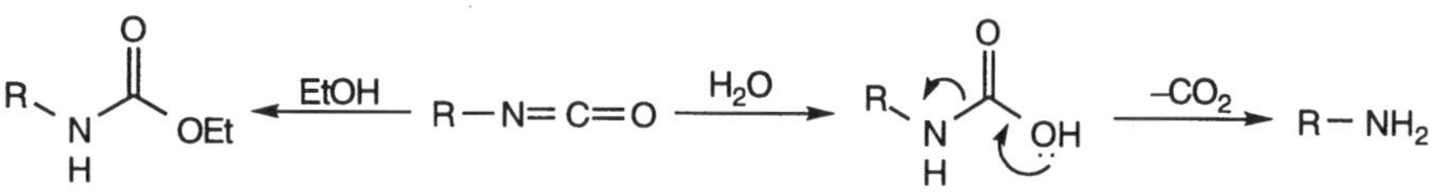

10.5 Summary

The N–N-containing functional groups undergo important interconversions that are of great synthetic value. The most important reactions are:

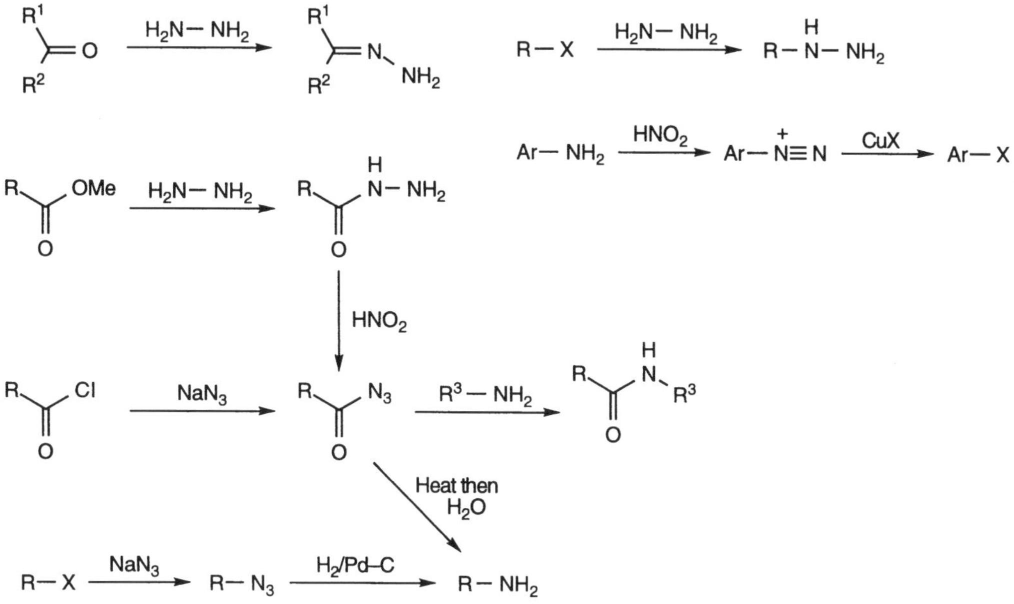

11 Oximes

Oximes are the first of the N–O-containing functional groups that we will consider. They are simple to make, and they undergo several important reactions.

11.1 Synthesis of oximes

Like ammonia, **hydroxylamine** and many hydrazine derivatives are basic and readily form salts such as:

$$HO-\overset{+}{N}H_3 \ \ \bar{C}l$$

These salts are more stable than their free amine counterparts (less easily oxidized) and so are the usual commercial source of these reagents.

By analogy with hydrazones, we can see that oximes can be made from ketones or aldehydes and hydroxylamine. Loss of water and conjugation of the oxime product drive the reaction forward. White crystalline oximes are often precipitated; their characteristic melting point is another classical way of identifying specific ketones and aldehydes.

e.g.

Importantly, the double bond of oximes is configurationally stable, so the above reaction yields a (separable) mixture of E- and Z-isomers:

E-isomer   *Z*-isomer

In some reactions of oximes this stereochemical feature can be a problem, but symmetrical ketones, of course, can only form a single stereoisomer.

11.2 Reactions of oximes

Hydrolysis

Oximes are relatively inert because the oxygen lone pair is conjugated with the π-bond. But vigorous hydrolytic conditions will cleave the C=N unit to regenerate the carbonyl compound:

Conjugation makes the C=N carbon less δ+ and so less susceptible to nucleophilic attack.

Dehydration of oximes

With oximes derived from **aldehydes** it is possible to eliminate H_2O, thereby generating the nitrile. Powerful dehydrating agents are required, and the mechanism requires the leaving group ability of OH to be improved. Phosphoric acid, PCl_5, and H_2SO_4 (conc.) are suitable reagents, but yields can be poor because of competition from the Beckmann rearrangement (next section).

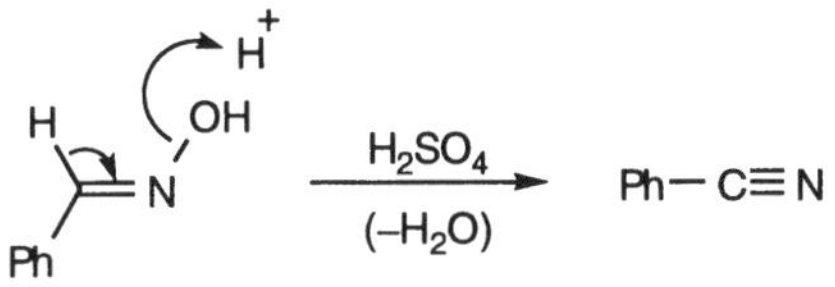

The Beckmann rearrangement

Oximes of ketones readily undergo the Beckmann rearrangement:

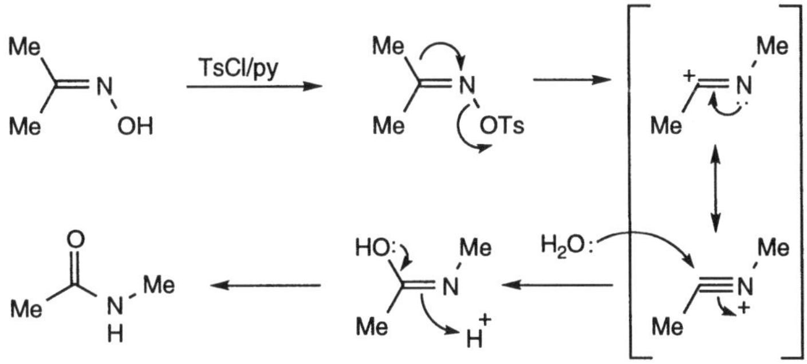

If an unsymmetrical oxime is used, only the alkyl/aryl group *anti* to the OH will migrate. This means that *E*- and *Z*-isomers of oximes will give different rearranged products. With oximes of aldehydes elimination competes with rearrangement.

The most important commercial use of the Beckmann rearrangement is the formation of ε-caprolactam, which is used in the nylon industry. In this case, the symmetrical cyclohexanone can give only one oxime, and the reaction is high yielding:

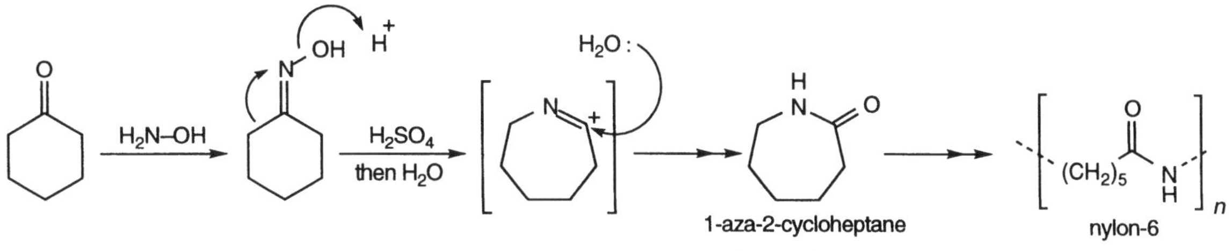

1-aza-2-cycloheptane
(ε-caprolactam)

nylon-6

Reduction

Hydrogenation over highly active catalysts leads to reduction of the π-bond **and** the N–O σ-bond, giving primary amines. This can be a useful synthetic procedure, although conditions are often quite vigorous.

e.g.

11.3 Summary

The key reactions are as follows:

12 *N*-Oxides

$$R_3\overset{+}{N}-\overset{-}{O}$$

12.1 Synthesis

Uniquely amongst the functional groups in this section, the N–heteroatom bond of *N*-oxides is **not** introduced intact in the synthesis of these compounds, but is delivered in an oxidation reaction.

e.g.

$$Et_3N \xrightarrow[\bullet\ H_2O_2/AcOH]{\bullet\ PhCO_2OH\ or} Et_3\overset{+}{N}-\overset{-}{O}$$

A peracid is the usual oxidant, either used directly or generated *in situ* from a carboxylic acid/hydrogen peroxide mixture. Almost all tertiary amines will undergo this reaction; primary and secondary amines usually form horrendous mixtures of products under these reaction conditions.

12.2 Reactions of *N*-oxides

N-Oxides are important for effecting *syn* eliminations. If the *N*-oxide possesses a β-hydrogen, the elimination proceeds smoothly on heating to about 100°C.

The *N*-oxide acts like a quaternary ammonium compound with an internal alkoxide base, so the (intramolecular) elimination is very easy. The reaction can also be considered as a six-electron rearrangement process—but whatever the mechanistic detail, the elimination cannot occur unless the β-hydrogen and the O⁻ are essentially *syn*-periplanar.

When more than one type of β-hydrogen is present, mixtures of elimination products can result. But if the other *N*-alkyl groups are methyls (which, of course, lack β-hydrogens) the elimination is regio- and stereo-specific.

e.g.

syn elimination

E-isomer

cf.

anti elimination

Z-isomer

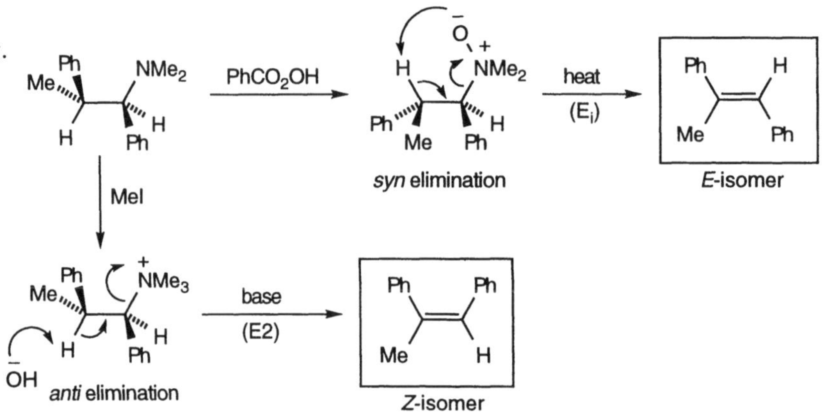

The above scheme shows the complementary nature of *syn* *N*-oxide elimination (E$_i$) and standard *anti* E2 elimination which is more common (e.g. with tetraalkyl ammonium compounds—see Chapter 2).

Whilst the trialkyl *N*-oxides are used for stereospecific elimination reactions, pyridine *N*-oxides (and other tertiary *N*-aromatics) are of value because they completely change the reactivity of the parent compound. The most notable example is perhaps pyridine itself:

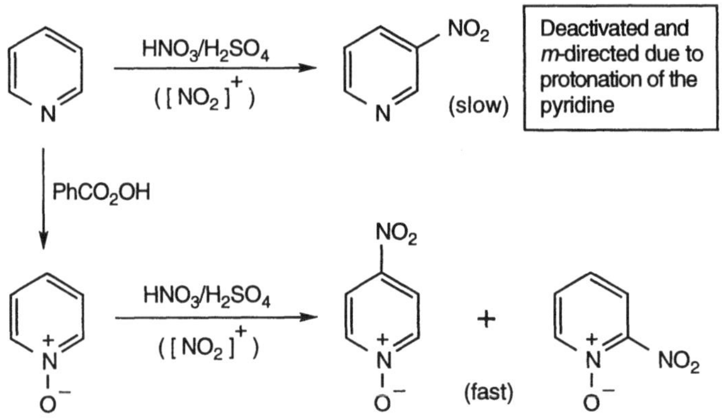

o,p-Attack of electrophiles on pyridine *N*-oxide
The change in reactivity and regioselectivity is due to the +M effect of the O$^-$, which stabilizes intermediates derived from *o,p*-attack. For *p*-attack:

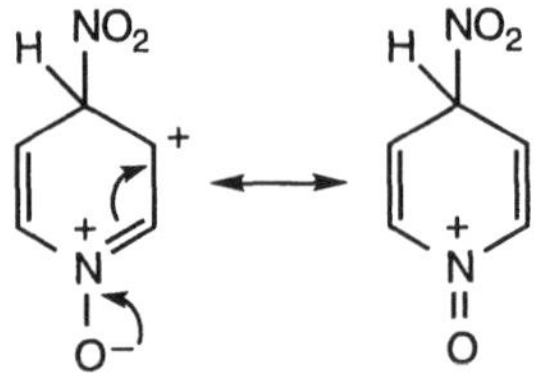

For *o*-attack:

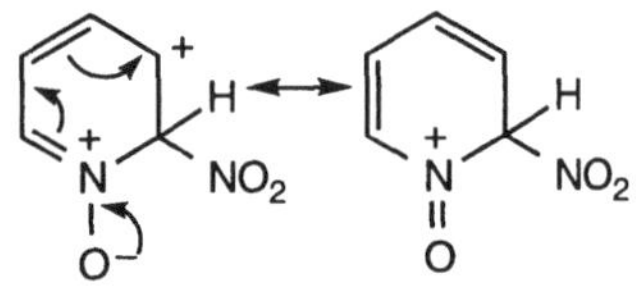

Similar tautomers (with a stable N=O double bond) are not possible for *m*-attack—try drawing the intermediate from *m*-attack to convince yourself of this.

This gives access to a different substitution pattern that is inaccessible directly from pyridine. When *N*-oxides are used in this way, it is important to be able to remove the oxygen afterwards, and this is simply accomplished by treatment with PCl$_3$.

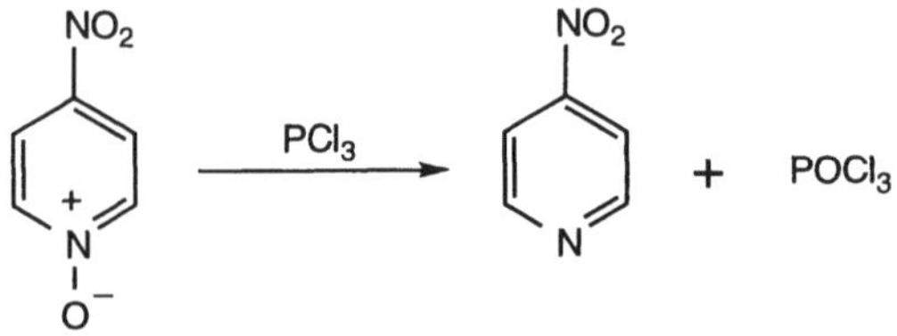

12.3 Summary

The key reactions are exemplified as follows:

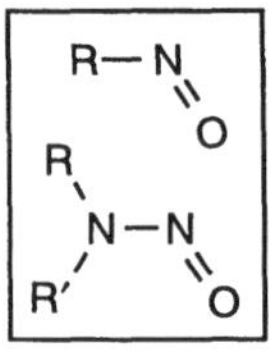

13 Nitroso compounds

Nitroso compounds are tautomeric with oximes, but oximes are thermodynamically more stable. Therefore, *C*-nitroso compounds are readily converted to the corresponding oxime if there is a hydrogen α to the nitroso group.

e.g.

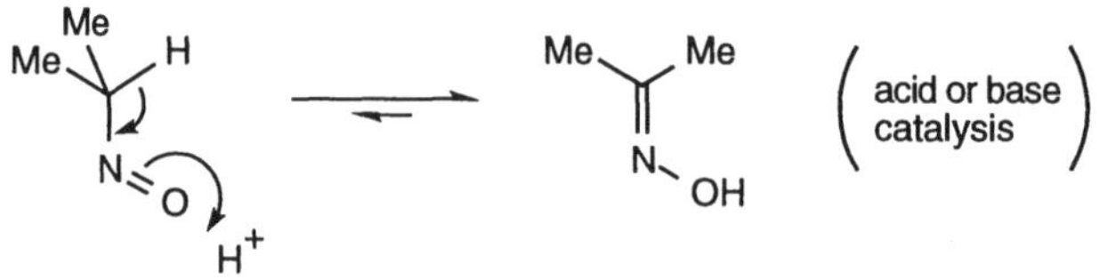

For this reason the chemistry of *C*-nitroso compounds is somewhat limited—but we will look at their synthesis and reactions, before discussing *N*-nitroso and *O*-nitroso compounds.

13.1 Aliphatic *C*-nitroso compounds

The synthesis of these compounds is not very easy—trapping of [NO]⁺ with a suitable nucleophile invariably leads to further reactions and degradation of the nitroso moiety. However, one type of nitroso compound can be obtained by treating ketones with NOCl.

e.g.

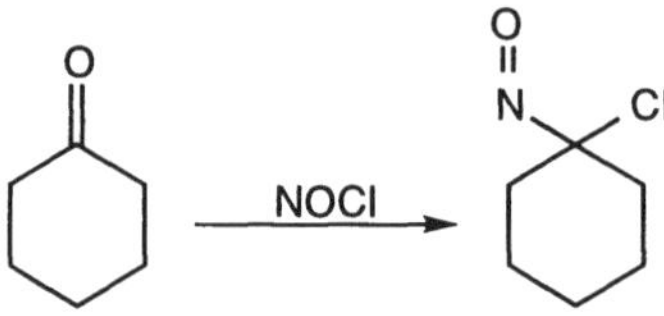

These α-chloronitroso compounds have been employed as dienophiles in Diels–Alder chemistry. Good stereocontrol in the cycloaddition can make this a valuable route to 4-hydroxyamines, as the N–O bond can be cleaved later by hydrogenation:

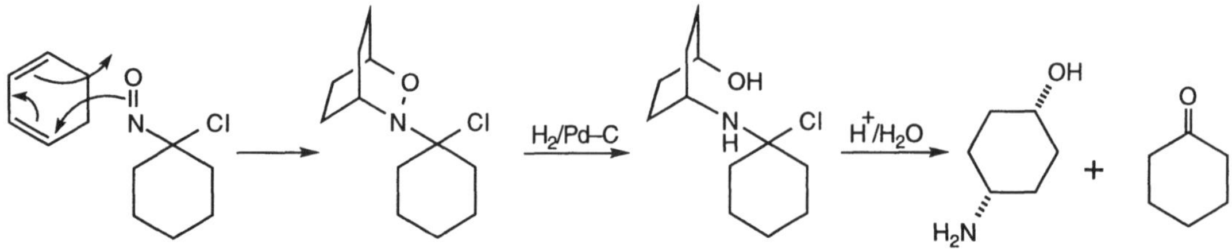

Acyl nitroso compounds are relatively easy to prepare, by oxidation of the hydroxamic acid. Again, their greatest use is in Diels–Alder chemistry, where the acyl nitroso compound can be made and trapped *in situ*:

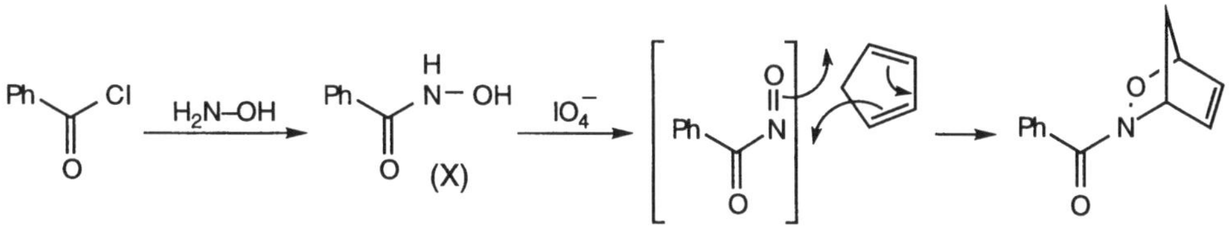

In the above scheme, X is a **hydroxamic acid** (RCONHOH). The *O,N*-dimethyl derivative reacts once with Grignard reagents to form ketones in high yields:

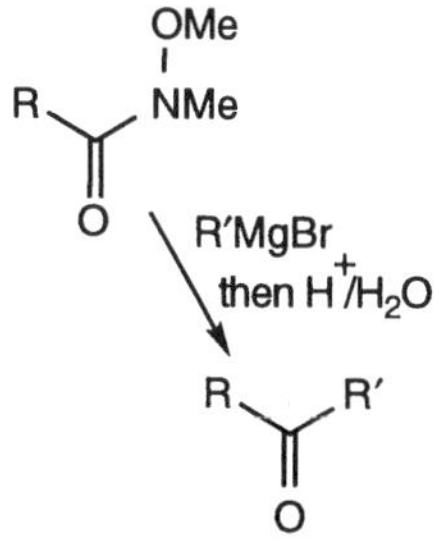

13.2 Aromatic *C*-nitroso compounds

Aromatic *C*-nitroso compounds cannot tautomerize to an oxime, and are further stabilized by conjugation. Thus with activated aromatic compounds, nitrous acid can be used to introduce the nitroso group. However, they undergo few important reactions: they are slowly oxidized to nitro compounds, or can form diazo compounds with aromatic amines:

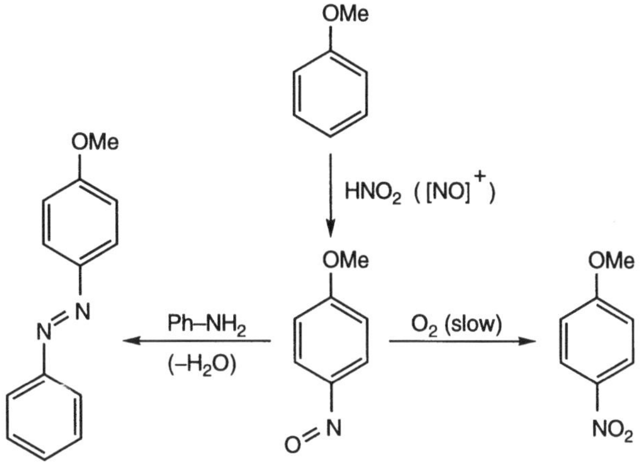

Q. Suggest a mechanism. Hint: the intermediate is a chelate of Mg^{2+}.

13.3 α-Diketones via nitroso compounds

The tautomerism of nitroso compounds can sometimes be exploited, as in the oxidation of ketones to α-diketones. This reaction utilizes the enol tautomer of the ketone as a carbon nucleophile that picks up $[NO]^+$.

e.g.

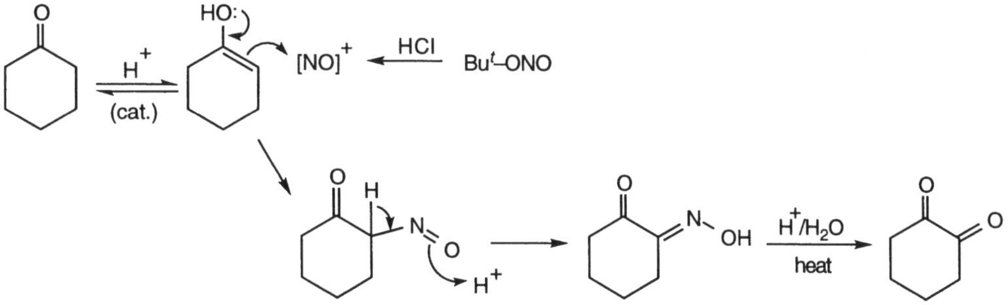

Here the acid catalyses enolization, causes $[NO]^+$ formation, and finally hydrolyses the oxime! The method is particularly valuable as it gives the reverse regiocontrol to selenium dioxide oxidation.

e.g.

13.4 *N*-Nitroso compounds

Carcinogenicity of *N*-nitroso compounds
Many *N*-nitroso compounds are potent carcinogens, because oxidation in the body generates highly electrophilic diazonium ions, which can alkylate DNA.

N-Nitroso compounds are of value in distinguishing primary, secondary, and tertiary amines. The test simply involves treating the amine with nitrous acid. Clearly, the $[NO]^+$ could be trapped by a nitrogen nucleophile, and for a secondary amine this leads to a simple *N*-nitroso compound—usually as an orange oil. Tertiary amines fail to react (or do so reversibly). Primary amines **do** react, but subsequent loss of water generates the aliphatic diazonium ion which loses nitrogen very easily.

e.g.

13.5 *O*-Nitroso compounds

Otherwise known as nitrite esters, these undergo just one important process—the Barton reaction. The formation and photolysis of an *O*-nitroso compound are summarized below:

The reaction is important because it can allow the functionalization of remote positions by H• abstraction—for example in the modification of steroids:

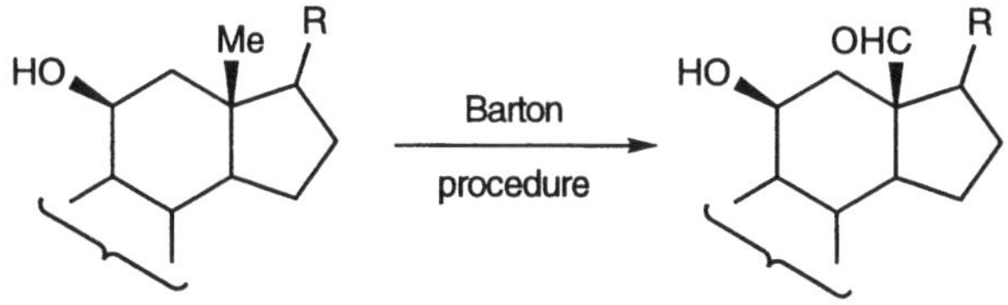

13.6 Summary

Aliphatic *C*-nitroso compounds tautomerize to the oxime if an α-hydrogen is present. The α-chloro and acyl *C*-nitroso compounds are useful in the Diels–Alder reaction:

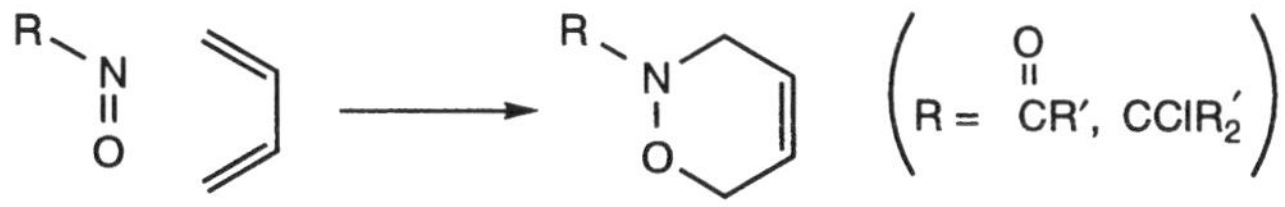

Using HCl/RONO, ketones can be converted to α-diketones via the nitroso/oxime tautomers.

Finally, HNO$_2$ is a classical test to differentiate between primary amines (N$_2$ evolved), secondary amines (orange *N*-nitroso compound formed), and tertiary amines (no reaction).

14 Nitro compounds

Nitro compounds play a central role in organonitrogen chemistry. The synthesis and properties of aromatic versus aliphatic nitro compounds are so different that we will consider them separately.

14.1 Aromatic nitro compounds

Synthesis

Perhaps one of the most infamous organonitrogen compounds is the explosive **trinitrotoluene** (TNT) obtained by vigorous nitration of toluene:

With few exceptions, aromatic nitro compounds are made by the direct nitration of aromatic compounds. The strength of the nitrating agent chosen depends on the reactivity of the aromatic compound. The most powerful nitrating agent is a mixture of concentrated nitric and sulphuric acids (so-called 'mixed acid'), which generates a high concentration of the very electrophilic nitronium ion:

$$2\,HNO_3 + H_2SO_4 \rightleftharpoons H_3O^+ + NO_3^- + HSO_4^- + [\,NO_2\,]^+$$
nitronium ion

This mixture will nitrate even quite strongly deactivated aromatics. e.g.

Crucially, nitration generates a strongly **deactivated** product (NO_2 being very electron withdrawing). Thus further nitration requires forcing conditions, such as refluxing in mixed acid.

The strongly acidic conditions for nitration always render aromatic amines inert—this is because the amine becomes protonated, changing the +M effect to a strong –I effect.

o,p-directing
activating

m-directing
deactivating

Protection of the amine by acylation gives an amide which is not protonated under the nitration conditions, but the nitrogen lone pair still has enough +M influence to strongly favour *o/p*-attack. The acyl protection can be removed by hydrolysis after the nitration step.

The standard mechanism for aromatic electrophilic substitution explains all of the above observations. Thus for activated aromatics *o/p*-attack is favoured:

In contrast, with deactivated aromatics, both *o*- and *p*-attack generate an intermediate in which one canonical form has adjacent positive charges. This does not occur for *meta*-attack which therefore predominates, although the –I effect slows down the reaction compared with benzene.

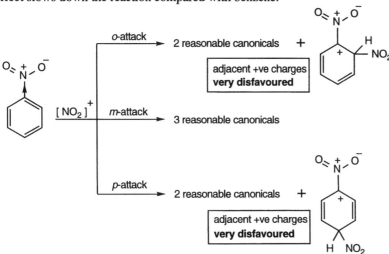

Q. Draw the three canonicals of the intermediate from *m*-attack, to prove that adjacent positive charges are not present.

Reactions of aromatic nitro compounds

There is just one important reaction—the reduction of nitro to amine. This can be effected by a wide range of reducing agents: Sn/HCl (old-fashioned, but still very effective!), H_2/Pd–C or $LiAlH_4$ are perhaps the commonest. This is the best way of introducing an **amino group** into an aromatic ring.

The nitro group can actually be completely removed from an aromatic ring via reduction to the amine followed by diazotization (with HNO_2), and then reductive loss of N_2; this latter step can be achieved using $NaBH_4$ (H^- displaces N_2), or more efficiently using an unusual hypophosphorous acid/Cu^+ mixture. We sometimes introduce 'NH_2', only to remove it subsequently, in order to exploit the powerful directing effect of the amino group. This can allow access to substitution patterns that would otherwise be very hard to obtain, and we can see this useful sequence of reactions in operation for the formation of 3-bromoaniline:

In this nitration the nitrogen *o,p*-directing effect overrides the weaker bromine *o,p*-directing effect

Lastly, the strongly electron withdrawing NO_2 group (–M and –I) can influence other functional group chemistry within an aromatic ring.

e.g.

This unusual nucleophilic aromatic substitution is only possible because the negative charge generated in the transition state can be delocalized onto the oxygens of the nitro groups:

These are two important canonical forms in which the incoming negative charge is delocalised onto the nitro oxygens.

There are three canonical forms where the incoming negative charge resides on the aromatic ring.

DNP
(see page 70)

In a similar manner, alkyl groups *ortho* or *para* to a nitro group can be quite readily deprotonated.

e.g.

However, this type of nitro-stabilized anion chemistry is really a feature of aliphatic nitro compounds (here the aromatic ring simply allows conjugation to extend the effect), and this chemistry really belongs in the next section.

14.2 Aliphatic nitro compounds

Synthesis

This is a very brief section because the nitro group is rarely introduced **into** aliphatic compounds...but simple nitro compounds are readily elaborated into more complex molecules. The usual building block is nitromethane, but other simple nitroalkanes are commercially available.

Reactions of aliphatic nitro compounds

Nitroalkanes are very useful in 'enolate' chemistry. They are remarkably acidic and analogy with carbonyl compounds is appropriate here. In nitroalkanes the delocalized negative charge is adjacent to the N^+, and this confers special stability on the anion.

Some other pK_a values are:

	8.8
HCN	9.1
PhOH	10.0
RNH_3^+	≈ 10
	≈ 11
NC⌣CN	11.2
MeOH	15.0

By looking at the pK_a values shown, we can say that the methoxide anion will deprotonate nitromethane but **not** acetone.

In contrast to enolates, nitro compounds have very little propensity to self-condense since (reversible) attack at N^+ does not generate stable products. Therefore crossed-condensations between nitroalkanes and aldehydes or ketones are easy to carry out. Subsequent dehydration (often during work-up) gives easy access to α,β-unsaturated nitro compounds.

In the α,β-unsaturated product, the nitro group can again act like a carbonyl group, making the system a Michael acceptor. So carbon nucleophiles can attack the end of the double bond.

With these two reactions available, it is easy to see how simple nitroalkanes can be quickly funtionalized to complex skeletons. However, only rarely is the nitro group required in the final target molecule. Two highly efficient functional group modifications can be used to transform the NO$_2$ group into a ketone or an amine.

Hydrolysis of nitro groups. Nitro groups can be hydrolysed directly under the vigorous conditions of the Nef reaction, provided there is an α-proton. The procedure requires trapping of the nitroenolate with aqueous H$_2$SO$_4$, and is effectively an imine hydrolysis.

e.g.

But reductive hydrolysis with TiCl$_3$(aq) is usually more efficient, generating the aldehyde or ketone.

e.g.

$$Et_2CH-NO_2 \xrightarrow{TiCl_3 \ (aq)} Et_2CO$$

The mechanism involves one-electron reductions, and can be summarized as:

Reduction of nitro groups. Many reducing agents will convert nitro groups into amines. The two commonest procedures use catalytic hydrogenation or $LiAlH_4$.

e.g.

So, the highly versatile nitro group can be converted into the amine moiety, one of the simplest but most important organonitrogen functional groups.

14.3 Summary

Aromatic nitro compounds are made by simple nitration of aryl compounds, and reduction gives aromatic amines.

e.g.

Aliphatic nitro compounds can be rapidly derivatized using carbonyl-like chemistry. Highly efficient 'hydrolysis' or reduction generates aldehydes/ketones or amines respectively.

e.g.

Further reading

Organonitrogen chemistry is an integral part of all mainstream organic textbooks. It also features strongly in books on synthesis, mechanism, reactivity, natural products, and more specific classes of compounds. Indeed, several aspects of organonitrogen chemistry from this book also feature in other OUP primers. However, here are six other books that you may find particularly interesting and helpful:

General (easy). J. McMurry (1992). *Organic Chemistry* (3rd ed.), Brooks/Cole, California. A very readable general organic textbook at a modest level—briefly covers almost all of the major topics in undergraduate chemistry.

General (harder). J. March (1992). *Advanced Organic Chemistry: Reactions, Mechanisms, and Structures* (4th ed.), Wiley, New York. More advanced general textbook—very good coverage of mainstream organic chemistry, with valuable literature references.

Synthesis. R. K. Mackie, D. M. Smith, and R. A. Aitken (1990). *Guidebook to Organic Synthesis* (2nd ed.), Longman, Harlow. There are dozens of books on synthesis—the role of nitrogen is well covered and easily located in this book.

Mechanism. P. Sykes (1986). *A Guidebook to Mechanism in Organic Chemistry* (6th ed.), Longman, Harlow. A classic, with lots of mechanisms involving nitrogen compounds.

Peptides. P. D. Bailey (1990/92). *An Introduction to Peptide Chemistry*, Wiley, Chichester. An overview of the topic, allowing you to see how the principles of organonitrogen chemistry apply to this important class of compounds. Many other excellent books on natural products are available, which give a wider perspective on the importance of nitrogen in biological systems.

Comprehensive. B. R. Brown (1994). *The Organic Chemistry of Aliphatic Nitrogen Compounds*, OUP, Oxford. A very thorough book of 767 pages, and this excludes aromatic nitrogen compounds.

Index